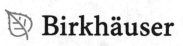

Pedro de M. Rios • Eldar Straume

Symbol Correspondences
for Spin Systems

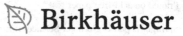

Pedro de M. Rios
Departamento de Matemática, ICMC
Universidade de São Paulo
São Carlos, SP
Brazil

Eldar Straume
Department of Mathematical Sciences
Norwegian University of Science
 and Technology
Trondheim
Norway

ISBN 978-3-319-35811-6   ISBN 978-3-319-08198-4 (eBook)
DOI 10.1007/978-3-319-08198-4
Springer Cham Heidelberg New York Dordrecht London

Mathematics Subject Classification (2010): 17B63, 22E70, 37J05, 53D99, 70G65, 70H05, 81-02, 81R05, 81R15, 81S30

Printed on acid-free paper

Springer is part of Springer Science+Business Media (www.springer.com)

Preface

This is a monograph intended for people interested in a more complete understanding of the relation between quantum and classical mechanics, here explored in the particular context of spin systems, that is, $SU(2)$-symmetric mechanical systems. As such, this book is aimed both at mathematicians with interest in dynamics and quantum theory and physicists with a mathematically oriented mind.

The mathematical prerequisites for reading this monograph are mostly elementary. On the one hand, some knowledge of Lie groups and their representations can be helpful, but since the group considered here is $SU(2)$, with its finite dimensional representations, not much more than lower-division college mathematics is actually required. In this sense, Chaps. 2 and 3 can almost be used as a particular introduction to some aspects of Lie group representation theory, for students of mathematical sciences. On the other hand, some knowledge of symplectic differential geometry can be helpful, but again, since the symplectic manifold considered here is the two-sphere with its standard area form, some of Chap. 4 can be seen as a concise presentation of the concepts of Hamiltonian vector field and Poisson algebra. As the remaining chapters of the book are built on these first three, the very diligent upper-division undergraduate student in mathematics, physics or engineering, with sufficient knowledge of calculus and linear algebra already acquired, should not find many serious difficulties in reading throughout most of this book.

Motivation is another story, however, and this monograph may not appeal to those not yet somewhat familiar with standard quantum mechanics and classical Hamiltonian mechanics. Rather, in fact, for those who have already taken a regular course in quantum mechanics with its standard treatment of angular momentum, some of the contents in Chap. 3 will look familiar, although most likely they will find the material covered in this chapter more precisely and explicitly treated than in other available texts. In particular, our detailed presentation of the rotationally invariant decomposition of the operator algebra of spin systems is not so easily found elsewhere. Similarly, a person with minimal knowledge of symplectic geometry and Hamiltonian dynamics may also find Chap. 4 too straightforward, and perhaps somewhat amusing when he looks at the detailed presentation of the rotationally invariant decomposition of the Poisson algebra on the two-sphere, but again, we were not able to find this last part in any other book.

Thus, this book is mainly addressing a person – student or researcher – who has pondered on the question of whether or how quantum (Heisenberg) dynamics and classical Hamiltonian dynamics (also called Poisson dynamics) can be precisely related. However, since this rather general question has already been asked and studied in various ways in many books and research papers, it is important to clarify how this question is addressed here and what is actually new in this book.

To this end, we must first emphasize that most ways of studying the above question are of two major types: the first one, proceeding from quantum dynamics to Poisson dynamics, and the second, proceeding in reverse order from Poisson dynamics to quantum dynamics, the latter also widely known as the process of "quantization" of a Poisson manifold. This monograph proceeds in the first way, which in its own turn can be approached according to two distinct methodologies: one which studies the so-called semiclassical limit directly from the quantum formalism, and another and more detailed approach, which first translates the quantum formalism to the classical formalism, in a process also referred to as "dequantization", and then studies the semiclassical limit in the dequantized setting. This more detailed methodology is the one we shall follow in the present monograph, albeit here we are limiting ourselves to the context of spin systems.

The advantage of working in the context of spin systems for this purpose is twofold: on the one hand, all quantum spaces are finite dimensional, which greatly simplifies the quantum formalism, and, on the other hand, since each process of dequantization depends on a choice of symbol correspondence, it is important to classify and make explicit all such possible choices, which is easier in the finite dimensional context, and this is fully done for the first time in Chap. 6.

The comparison of all such choices of symbol correspondences is continued in Chap. 7, which presents a detailed study of spherical symbol products, and Chap. 8, which starts the study of their asymptotic limit of high spin numbers. Together, Chaps. 6–8 of this book present for the first time in the literature a systematic study of general symbol correspondences for spin systems, with their symbol products and asymptotic limit of high spin numbers. In so doing, as far as we know these chapters also present the first systematic study, in books as well as in research papers, of how classical mechanics may or may not emerge as an asymptotic limit of quantum mechanics, in a rather simple and precise context.

Contents

Chapter 1
Introduction

Non-relativistic quantum mechanics was formulated in the first half of the twentieth century by various people, most prominently by Heisenberg [38] and Schrödinger [66] who independently followed, in the mid 1920s, on the preliminary work of Bohr [17]. The latter attempted to modify non-relativistic classical mechanics, as introduced by Newton in the seventeenth century and developed in the following two centuries by Euler, Lagrange, Hamilton, Jacobi and Poisson, among others,[1] in order to be compatible with the energy and momentum quantization postulates presented first by Planck and then by Einstein and de Broglie in the 1900s.

While the pioneering work of Heisenberg with the help of Born and Jordan introduced what at first became known as "matrix mechanics" [19, 20], the work of Schrödinger produced a partial differential equation for the "wave function". Although Schrödinger was soon able to link his approach to Heisenberg's, the two approaches were fully brought together a little later by von Neumann [76, 78]. His final formalism for non-relativistic quantum mechanics is based on the concepts of vectors and operators on a complex Hilbert space. In many cases these are infinite dimensional spaces and, in fact, the complete formalism had to expand on these concepts in order to accommodate distributions and functions which are not square integrable, thus leading to Gelfand's rigged Hilbert spaces [31], quite later.

From the start, Bohr emphasized the importance of relating the measurable quantities in quantum mechanics to the measurable quantities in classical mechanics. However, his so-called "correspondence principle" was not so easily implemented at the level of relating the two mathematical formalisms in a coherent way. At first, the basic mathematical concepts for describing classical conservative dynamics, functions on a phase space of positions and momenta (a symplectic affine space), were brought in a very contrived way to the quantum formalism through a series of cooking recipes called "quantization". Conversely, the "classical limit" of a quantum

---

[1]Poincaré, for instance, came a little later and his name is perhaps better associated to two other contemporary revolutions in dynamics: chaos and relativity.

© Springer International Publishing Switzerland 2014
P. de M. Rios, E. Straume, *Symbol Correspondences for Spin Systems*,
DOI 10.1007/978-3-319-08198-4_1

dynamical system, where classical dynamics should prevail, is often a singular limit and in the initial formulations of quantum mechanics it is not the phase space, but rather either the space of positions or the space of momenta, that is present in a more explicit way.

Some of these problems were addressed by Dirac [25, 26] already in his PhD thesis, but a clearer approach for relating the classical and the quantum mathematical formalisms was first introduced by Weyl [85] via a so-called symbol map, or symbol correspondence from bounded operators on Hilbert space $L^2_{\mathbb{C}}(\mathbb{R}^n)$ to functions on phase space $\mathbb{R}^{2n}$, these latter functions depending on Planck's constant $\hbar$. Soon after, Wigner [90] expanded on Weyl's idea to produce an $\hbar$-dependent function $\mathcal{W}_{\psi} \in L^1_{\mathbb{R}}(\mathbb{R}^{2n})$ as the phase-space representation of a wave function $\psi \in L^2_{\mathbb{C}}(\mathbb{R}^n)$. While real and integrating to 1 over phase space, $\mathcal{W}_{\psi}$ is only a pseudo probability distribution on $\mathbb{R}^{2n}$ because it can take on negative values, as opposed to $|\psi|^2$ which is a true probability distribution on $\mathbb{R}^n$.

Despite this shortcoming, Wigner's function inspired Moyal [50] to develop an alternative formulation of quantum mechanics as a "phase-space statistical theory", following on Weyl's correspondence, where the Poisson bracket of functions on $\mathbb{R}^{2n}$ was replaced by an $\hbar$-dependent bracket of $\hbar$-dependent functions on $\mathbb{R}^{2n}$, whose "classical limit" is the Poisson bracket. Moyal's skew-symmetric bracket still satisfies Jacobi's identity and, in fact, can be seen as the commutator of an $\hbar$-dependent associative product on the space of $\hbar$-dependent Weyl symbols. Although Moyal's product is written in terms of bi-differential operators, an integral formulation of this product had been previously developed, first by von Neumann [77] soon after Weyl's work, then re-discovered by Groenewold [35]. This Weyl-Moyal approach was further developed by Hörmander [27,39,40], among others, into the calculus of pseudo-differential and Fourier-integral operators. It also inspired the deformation quantization approach started by Bayen, Flato, Frondsal, Lichnerowicz and Sternheimer for general Poisson manifolds [9].

However, while the Hilbert space used to describe the dynamics of particles in 3-dimensional configuration-space of positions is generally infinite dimensional, if one restricts attention to rotations around a point, only, the corresponding Hilbert space is finite dimensional. Historically, the necessity to introduce an independent, or intrinsic finite dimensional Hilbert space for studying the dynamics of atomic and subatomic particles stemmed from the understanding that a particle such as an electron has an intrinsic "spinning" which is independent of its dynamics in 3-space. An extra degree of freedom, therefore.

After the hypothesis of an extra degree of freedom was first posed in 1924 by Pauli [52] and 1 year later identified by Goudsmit and Uhlenbeck [71, 72] as an intrinsic "electron spin", the spin theory was developed around 1930, mostly by Wigner [88, 89, 91], but also by Weyl [86], through a careful study of the group of rotations and its simply-connected double cover $SU(2)$, and their representations.

Because this intrinsic quantum dynamics in a finite-dimensional Hilbert space had no obvious classical counterpart, at that time, the necessity to relate it with a corresponding classical dynamics was not present from the start. In fact, to the extra degree of freedom of spin, there should correspond a 2-dimensional phase space and, by $SU(2)$ invariance, this phase space must be the homogeneous 2-sphere S^2. Thus, the fact that its classical configuration-space cannot be products of euclidean 3-space or the real 2-sphere was at first interpreted as an indication that a physical correspondence principle for spin systems was impossible.

This may explain why the mathematical formulation of a Weyl's style correspondence for spin systems was developed much later. Thus, while Weyl's to Moyal's works are dated from the mid 1920s to the late 1940s, an incomplete version of a symbol correspondence for spin systems was first set forth in mid 1950s by Stratonovich [67] and a first more complete version was presented in mid 1970s by Berezin [10–12]. Then, in the late 1980s, the work of Varilly and Gracia-Bondia [73] finally completed and expanded the draft of Stratonovich. Its contemporary work of Wildberger [92] was followed by the relevant works of Madore [47], in the 1990s, and of Freidel and Krasnov [30], in early 2000s. All these works produced a very good, but in our view, still incomplete understanding of symbol correspondences and symbol products, for spin systems.

From the mathematical point of view, such a late historical development is surprising because quantum mechanics in finite-dimensional Hilbert space, on top of being far closer to Heisenberg's original matrix mechanics, is far simpler than quantum mechanics in infinite-dimensional Hilbert space. On the other hand, the corresponding phase space for the classical dynamics of spin systems, the 2-sphere S^2, has nontrivial topology, as opposed to $\mathbb{R}^{2n}$. Nonetheless, one should expect that a mathematical formulation of Bohr's correspondence principle "à la Weyl" would have been developed for spin systems in a more systematic and complete way than it had been achieved for mechanics of particles in k-dimensional Euclidean configuration space. However, despite the many developments in the last 50 years, such a systematic mathematical formulation of correspondences in spin systems was still lacking. In particular, we wanted a solid mathematical presentation unifying the main scattered papers on the subject (some of them based also on heuristic arguments) and completing the various gaps, so as to clarify the whole landscape and open new paths to still unexplored territories.

Here we attempt to fulfill this goal, at least to some extent. This monograph, written to be as self-contained as possible, is organized as follows.

In Chap. 2 we review generalities on Lie groups and their representations, with special attention to the groups $SU(2)$ and $SO(3)$.

In Chap. 3 we present the elements for quantum dynamics of spin systems, that is, $SU(2)$-symmetric quantum mechanical systems. First, we carefully review its representation theory which defines spin-j systems and their standard basis of Hilbert space. Then, after reviewing the tensor product and presenting the space of operators, with its irreducible summands and standard coupled basis, we recall how the operator product is related to quantum dynamics via Heisenberg's equation involving the commutator and then we present in a very detailed way the $SO(3)$-invariant decomposition of the operator product, its multiplication rule in the coupled basis of operators, with its parity property.

Almost all choices, conventions and notations used in this chapter are standard ones used in the vast mathematics and physics literature on this subject. In particular, we introduce and manipulate Clebsch-Gordan coefficients and the various Wigner symbols in a traditional way, without resorting to the more modern language of "spin networks" which, in our understanding, would require the introduction of additional definitions, etc., which nonetheless would not contribute significantly to the main results to be used in the rest of the book.

In Chap. 4 we present the elements of the classical $SU(2)$-symmetric mechanical system. After reviewing some basic facts about the 2-sphere as a symplectic manifold and presenting the key concepts of classical Hamiltonian dynamics and Poisson algebra of smooth functions on S^2, defining the classical spin system, we present in some detail the space of polynomial functions on S^2, the spherical harmonics, followed by a detailed presentation of the $SO(3)$-invariant decomposition of the pointwise product and the Poisson bracket of functions on S^2.

Chapter 5, Intermission, pauses the study of spin systems. Here we present a brief historical overview of symbol correspondences in affine mechanical systems, in preparation for the remaining and most novel chapters of the book.

Thus, in Chap. 6 we present the $SO(3)$-equivariant symbol correspondences between operators on a finite-dimensional Hilbert space and (polynomial) functions on S^2. After defining general symbol correspondences and determining their moduli space, we distinguish the isometric ones, the so-called Stratonovich-Weyl symbol correspondences. Then, we present explicit constructions of general symbol correspondences, particularly the ones via coupled-standard basis and via an operator kernel, introducing the key concept of characteristic numbers of a symbol correspondence, and the general covariant-contravariant duality for non-isometric ones. Besides correspondences of Stratonovich-Weyl type, special attention is devoted to the non-isometric correspondence defined by Berezin via Hermitian metric, which is then generalized to correspondences defined via coherent states.

In Chap. 7 we study in detail the products of symbols induced from the operator product via symbol correspondences, the so-called twisted products. After definition and basic properties, we produce explicit expressions for some twisted products of cartesian symbols, valid for all $n = 2j \in \mathbb{N}$. Then, we describe the formulae for general twisted products of spherical harmonics, Y_l^m, discussing some of their common properties. Next, we present a detailed study of integral trikernels, which define twisted products via integral equations. We state the general properties and produce various explicit formulae (including some integral ones) for these trikernels. In so doing, we arrive at formulae for various functional transforms that generalize the Berezin transform and we also see that it is not so easy to infer a simple closed formula for the Stratonovich trikernel in terms of midpoint triangles (in a form first inquired by Weinstein [83] based on the analogy with the Groenewold-vonNeumann trikernel) unless, perhaps, asymptotically.

Then, in Chap. 8 we start the study of the asymptotic $j \to \infty$ limit of symbol correspondence sequences and their sequences of twisted products. In this monograph, we focus on the high-j asymptotics for finite l, here called low-l high-j-asymptotics, leaving high-l-asymptotics to a later opportunity. We show that

Poisson (anti-Poisson) dynamics emerge in the asymptotic $j \to \infty$ limit of the standard (alternate) Stratonovich twisted product as well as the standard (alternate) Berezin twisted product, but this is not the generic case for sequences of twisted products, not even in the restricted subclass of twisted products induced from isometric correspondence sequences. Thus, we characterize some kinds of symbol correspondence sequences based on their asymptotic properties and also discuss some measurable consequences.

In Chap. 9 we present some concluding thoughts. For spin systems, adding to Rieffel's old theorem on $SO(3)$-invariant strict deformation quantizations of the 2-sphere [56], one now also has to take into consideration the fact that generic symbol correspondence sequences do not yield Poisson dynamics in the asymptotic limit of high spin numbers. In light of these results, old and new, we reflect on the peculiar nature of the classical-quantum correspondence.

Finally, in the Appendix we gather proofs of some of the propositions and a theorem, which were stated in the main text.

Acknowledgements During work on this project, we have benefited from several mutual visits. We thank FAPESP (scientific sponsor for the state of São Paulo, Brazil) and USP, as well as NTNU and the Norwegian NSF, for support of these visits. We are also grateful to UC Berkeley's Math. Dept. for hospitality during some of the periods when we were working on this project. Again, we thank the above sponsors for financial support of these stays. Many of the original results in this monograph were first presented at the conference "Geometry and Algebra of PDE's", Tromsø, 27–31 August 2012. We thank the organizers for the opportunity and some of the people in the audience for the interest, which stimulated us to wrap up this work in its present form. We also thank, in particular, Robert Littlejohn and Austin Hedeman for discussions, Marc Rieffel for taking time to read and comment on some parts of the monograph, as well as posing interesting questions and remarks that led to an improved text, and Nazira Harb for collaboration in Appendix 10.2 and for sharing with us some of her results in Examples 7.2.13 and 7.2.14. Finally, we thank Alan Weinstein for invaluable suggestions on a preliminary version of this monograph, dated December 2012. Since then, we've also benefited from various interesting and important suggestions and comments from the anonymous reviewers, to whom we are particularly grateful.

Chapter 2
Preliminaries

This chapter presents basic material on Lie groups and their representations, with emphasis on the Lie groups $SO(3)$ and $SU(2)$, as a preparation for next chapters. The reader already very familiar with the subject may skip or just glance through this chapter. For the reader unfamiliar or vaguely familiar with the subject, we refer to [2, 6, 28, 32, 41, 69, 75], for instance, for more detailed presentations.

2.1 On Lie Groups and Their Representations

The notion of "symmetry", expressed mathematically in terms of groups of transformations, plays a fundamental role in classical as well as quantum mechanics.[1]

Recall that a group is a set G together with a map $\mu : G \times G \to G$, $(g_1, g_2) \mapsto \mu(g_1, g_2) = g_1 g_2$, satisfying

$$(i)\ (g_1 g_2) g_3 = g_1 (g_2 g_3), \forall g_1, g_2, g_3 \in G,$$

$$(ii)\ \exists e \in G \text{ s.t. } eg = ge = g,\ \forall g \in G,$$

$$(iii)\ \forall g \in G, \exists g^{-1} \in G \text{ s.t. } gg^{-1} = g^{-1}g = e.$$

If the set G is a smooth manifold and the map $\Lambda : G \times G \to G$, $(g_1, g_2) \mapsto g_1^{-1} g_2$ is smooth, then G is called a *Lie group*. The concept of subgroup is the natural one, so, $H \subset G$ is a *subgroup* of G if $g_1^{-1} g_2 \in H$, $\forall g_1, g_2 \in H$.

Recall also that given two groups G, H, a group homomorphism from G to H is a map $\Psi : G \to H$ satisfying $\Psi(g_1 g_2) = \Psi(g_1)\Psi(g_2)$. If G, H are Lie groups and

---

[1]A more general but in some respects weaker algebraic structure that can be used to describe partial symmetries is that of a *groupoid* (cf. [84]). We shall meet a groupoid that plays an important role in the context of this monograph later on, cf. Definition 7.1.19.

© Springer International Publishing Switzerland 2014   7
P. de M. Rios, E. Straume, *Symbol Correspondences for Spin Systems*,
DOI 10.1007/978-3-319-08198-4_2

Ψ is smooth, then it is a Lie group homomorphism. If Ψ is also bijective, then it is a (Lie) group isomorphism.

Now, we say that a group G acts (from the left) on a set M if there is a map $\Phi : G \times M \to M$, such that

$$(i)\ em = m \quad \text{and} \quad (ii)\ g_2(g_1 m) = (g_2 g_1)m, \quad \text{for all } m \in M, \ g_i \in G,$$

where we have used the simpler notation gm for $\Phi(g, m)$.[2] In other words, there is a group homomorphism

$$\bar{\Phi} : G \to S(M), \ \bar{\Phi}(g) : m \mapsto gm, \tag{2.1}$$

where $S(M)$ is (a subgroup of) the group of invertible maps $M \to M$.

Associated with the above action of G on M there is the induced (left) action on the set of scalar functions on M:

$$g \in G : f \to f^g, \ f^g(m) = f(g^{-1}m). \tag{2.2}$$

In the typical applications in geometry and physics, G is a Lie group, M is a smooth manifold, and the above maps and functions are smooth. M might also have some specific algebraic or geometric structures and $S(M)$ is the group (or a subgroup) of transformations preserving (some or all of) the relevant structures.

2.1.1 Classical Groups over the Classical Fields

Many of the familiar examples of Lie groups arise from considering *matrix groups*, which are subgroups of $GL_{\mathbb{K}}(n)$, the group of invertible matrices over $\mathbb{K} = \mathbb{R}, \mathbb{C}$, or $\mathbb{H}$ (quaternions), which is isomorphic to the group $GL(V)$ of invertible linear transformations of an n-dimensional $\mathbb{K}$-vector space $M = V \simeq \mathbb{K}^n$, via the usual left action of matrices on column vectors. Note, however, in the case $\mathbb{K} = \mathbb{H}$ scalar multiplication must act on the right side of column vectors to ensure that matrix multiplication from the left side is an $\mathbb{H}$- linear operator. $\mathbb{K}^n$ has the usual Hermitian inner product $\langle u, v \rangle = \sum_{j=1}^{n} \bar{u}_j v_j$ and norm $\|u\| = \langle u, u \rangle^{1/2}$.

The natural extension of scalars, from real to complex to quaternion, has several important consequences for representations of groups. First of all, for $n \geq 1$ there are natural inclusion maps c, q and linear isomorphisms r, c'

$$\mathbb{R}^n \to^c \mathbb{C}^n = \mathbb{R}^n + i\mathbb{R}^n \to^r \mathbb{R}^n \oplus \mathbb{R}^n, \tag{2.3}$$

$$\mathbb{C}^n \to^q \mathbb{H}^n = \mathbb{C}^n + j\mathbb{C}^n \to^{c'} \mathbb{C}^n \oplus \mathbb{C}^n,$$

---

[2]Of course, there is a similar definition of an action from the right.

where c and r (resp. q and c') are $\mathbb{R}$-linear (resp. $\mathbb{C}$-linear) maps.[3] Then there are corresponding injective homomorphisms of groups

$$GL_{\mathbb{R}}(n) \to^c GL_{\mathbb{C}}(n) \to^r GL_{\mathbb{R}}(2n), \qquad (2.4)$$

$$GL_{\mathbb{C}}(n) \to^q GL_{\mathbb{H}}(n) \to^{c'} GL_{\mathbb{C}}(2n).$$

Again, c, q in (2.4) denote inclusions, whereas r and c' replace each entry of a matrix by a 2-block, as follows:

$$r : (x + iy) \to \begin{pmatrix} x & -y \\ y & x \end{pmatrix}, \quad c\prime : (z_1 + jz_2) \to \begin{pmatrix} z_1 & -\bar{z}_2 \\ z_2 & \bar{z}_1 \end{pmatrix}. \qquad (2.5)$$

The terminology explained in the last footnote also makes sense for the above groups and homomorphisms.

Of particular interest to us is the unitary group $U(V) \subset GL(V)$ consisting of operators preserving the norm of vectors. Its conjugacy class consists, in fact, of all maximal compact subgroups of $GL(V)$, as follows directly from Lemma 2.1.1 (i) below. In terms of matrices, the $\mathbb{K}$-unitary group $U_{\mathbb{K}}(n) \subset GL_{\mathbb{K}}(n)$ consists of the matrices A whose inverse A^{-1} is the adjoint $A^* = \bar{A}^T$ (conjugate transpose), or equivalently, $\|Au\| = \|u\|$ for all vectors u. For the three cases of scalar field $\mathbb{K}$, these are the compact classical groups

$$O(n) \subset U(n) \subset Sp(n) \qquad (2.6)$$

called the orthogonal, unitary, and symplectic group,[4] respectively. The special orthogonal and special unitary groups $SO(n) \subset SU(n)$ are constrained by the additional condition $\det(A) = 1$. Observe that the map r in (2.5) yields an isomorphism $U(1) \simeq SO(2)$ by restriction to $x^2 + y^2 = 1$, whereas the map c' yields an isomorphism $Sp(1) \simeq SU(2)$ by restriction to $|z_1|^2 + |z_2|^2 = 1$.

In fact, these are particular cases of the following observation: since at each "doubling procedure" $\mathbb{R}^n \to \mathbb{C}^n \to \mathbb{H}^n$ a new algebraic structure is introduced, which needs to be preserved by the symmetry group, we also have the inclusions

$$Sp_{\mathbb{H}}(n) \equiv Sp(n) \subset U(2n) \equiv U_{\mathbb{C}}(2n),$$

$$U_{\mathbb{C}}(n) \equiv U(n) \subset O(2n) \equiv O_{\mathbb{R}}(2n).$$

---

[3] The mappings c, q, r, c' are generally referred to as complexification, quaternionification, real-reduction, and complex-reduction, respectively.

[4] The reader should not confuse the symplectic group $Sp(n) \equiv Sp_{\mathbb{H}}(n)$, which is a compact group over the quaternions, with the group $Sp_{\mathbb{R}}(2n) \subset GL_{\mathbb{R}}(2n)$ which is a noncompact group over the reals, often also called the symplectic group, in short for the *real linear symplectic group* – we will briefly mention a few basic things about this latter in the intermission, Chap. 5.

2.1.2   Linear Representations of a Group

Let us recall the basic definitions and results about finite dimensional linear representations of a group G. Namely, a *representation* of a (Lie) group G on $V \simeq \mathbb{K}^n$ is a (Lie) group homomorphism

$$\varphi : G \to GL(V) \simeq GL_{\mathbb{K}}(n). \tag{2.7}$$

We also say that φ is a $\mathbb{K}$-representation to emphasize that G acts on V by $\mathbb{K}$-linear transformations. Recall from elementary linear algebra that an isomorphism of the groups $GL(V)$ and $GL_{\mathbb{K}}(n)$, as indicated in (2.7), amounts to the choice of a basis for V. The representation is said to be *orthogonal*, *unitary*, or *symplectic* if the image $\varphi(G)$, viewed as a subgroup of $GL_{\mathbb{K}}(n)$, lies in the corresponding unitary group (2.6).

Two representations of G on vector spaces V_1 and V_2 are said to be equivalent (or isomorphic) if there is a G-equivariant linear isomorphism $F : V_1 \to V_2$, namely satisfying $F(gv) = gF(v)$ for all $g \in G$, $v \in V_1$. In terms of matrices, if $\varphi_i : G \to GL_{\mathbb{K}}(n), i = 1, 2$, are two matrix representations, then their equivalence $' \simeq '$ (as $\mathbb{K}$-representations) is defined with respect to some fixed $A \in GL_{\mathbb{K}}(n)$, as follows:

$$\varphi_1 \simeq \varphi_2 \iff \varphi_2(g) = A\varphi_1(g)A^{-1}, \forall g \in G. \tag{2.8}$$

However, the shorthand $\varphi_1 = \varphi_2$ is often used to mean $\varphi_1 \simeq \varphi_2$.

For simplicity, a representation φ of G on V is sometimes denoted by the pair (G, V), with φ tacitly understood. A subspace $U \subset V$ is said to be G-invariant if $gv \in U$ for each $v \in U$, and the representation (2.7) is *irreducible* if there is no G-invariant subspace strictly between $\{0\}$ and V. In representation theory, the classification of all irreducible representations of G, up to equivalence, is a central problem. More generally, the $\mathbb{K}$-representation (G, V) is said to be *completely reducible* if V decomposes into a direct sum $V = V_1 \oplus V_2 \oplus \ldots \oplus V_k$ of irreducible G-invariant $\mathbb{K}$-subspaces V_i, each defining an irreducible representation (G, V_i) with homomorphism $\varphi_i : G \to GL(V_i)$.

Henceforth, we shall assume the group G is compact, which in view of the following lemma simplifies the representation theory considerably.

Lemma 2.1.1. *For a compact group G the following hold:*

(i) *All $\mathbb{K}$-representations φ are $\mathbb{K}$-unitary, namely $\varphi : G \to U(V)$ for a suitable Hermitian inner product on V.*
(ii) *All representations are completely reducible.*

The standard proof is often referred to as "Weyl's unitary trick", using the fact that (G, V) has a G-invariant Hermitian inner product $\langle , \rangle$. It is found by averaging a given inner product $(,)$ over G, namely we set

$$\langle u, v \rangle = \int_G (gu, gv) dg. \tag{2.9}$$

Here dg denotes the (normalized) Haar measure on G, which is bi-invariant in the sense that the functions $f(g), f(kg), f(gk)$ on G have the same integral, for $k \in G$ fixed. Now property (ii) follows from the observation that for any G-invariant subspace $U \subset V$, the orthogonal complement $U^\perp$ is also G-invariant.

From complete reducibility one can show that any representation has a unique decomposition $\varphi = \sum \varphi_i$ into a (internal) direct sum of irreducibles, often written as an ordinary sum $\varphi = \varphi_1 + \varphi_2 + \ldots + \varphi_k$. Some of the φ_i may be equivalent, of course, so the notation such as $\varphi = 2\varphi_1 + 3\varphi_2$ should be clear enough.

Note, however, that the internal splitting (or direct sum) as explained above is different from the notion of an external splitting of a representation (G, V), or equivalently an external direct sum of representations (G_i, V_i). Namely, G is the product of the groups G_i and V is the direct sum of the V_i, each G_i acting nontrivially only on the summand V_i. For example, with two factors we have

$$(G, V) = (G_1 \times G_2, V_1 \oplus V_2) = (G_1, V_1) \oplus (G_2, V_2) , \quad \varphi = \varphi_1 \oplus \varphi_2.$$

Moreover, in the case $G_1 = G_2 = H$ so that $G = H \times H$, restriction of the above external splitting to the diagonal subgroup $\Delta H \simeq H$ of G yields an internal splitting $\varphi|_{\Delta H} = \varphi_1 + \varphi_2$ of the representation (H, V).

Clearly, an irreducible representation may become reducible by extension of the scalar field $\mathbb{K}$. For example, a real irreducible representation ψ may split after complexification,

$$c\psi : G \to^\psi GL_\mathbb{R}(n) \to^c GL_\mathbb{C}(n), \tag{2.10}$$

say $c\psi = \varphi_1 + \varphi_2$. As an example, consider the standard representation μ_1 of the circle group $U(1) = \{z \in \mathbb{C}; |z| = 1\}$ acting by scalar multiplication on the complex line $\mathbb{C}$. Realification of μ_1 yields the standard representation ρ_2 of $SO(2) \simeq U(1)$, acting by rotations on $\mathbb{R}^2$. Next, complexification of ρ_2 means regarding the rotation matrices as elements of $GL_\mathbb{C}(2)$, and here they can be diagonalized simultaneous. Refering to (2.4) we illustrate the effect of the two "operations" r and c as follows:

$$(x + iy) \to^r \begin{pmatrix} x & -y \\ y & x \end{pmatrix} \to^c \begin{pmatrix} x + iy & 0 \\ 0 & x - iy \end{pmatrix}, \mu_1 \to \rho_2 \to \mu_1 + \bar{\mu}_1 \tag{2.11}$$

Here, in the last step we have changed the basis of $\mathbb{C}^2$ so that the matrices become diagonal. Thus, in particular, the irreducible representation ρ_2 becomes reducible when complexified.

Next, note that conjugation of complex matrices yields a group isomorphism

$$t : GL_\mathbb{C}(n) \to GL_\mathbb{C}(n), \quad A \to \bar{A}.$$

Therefore, a $\mathbb{C}$-representation φ composed with t is the *complex conjugate* representation $t\varphi = \bar{\varphi} : g \to \overline{\varphi(g)}$; for example see (2.11). In a similar vein, there is a closely related construction, namely the *dual* (or contragradient) representation φ^T of φ, acting on the dual space $V^* = Hom(V, \mathbb{C})$ and defined by

$$\varphi^T(g) = \varphi(g^{-1})^T, \tag{2.12}$$

where Φ^T denotes the dual of an operator Φ on V. However, since G is compact it follows that the dual representation is the same as the complex conjugate representation. In fact, in terms of matrices with respect to dual bases in V and V^*, for $\varphi(g)$ unitary the left side of (2.12) is the matrix $\bar{\varphi}(g)$, hence $\varphi^T = \bar{\varphi}$.

Together with the homomorphisms in (2.4) we now have the six "operations" $r, c, c', q, t, 1$, where 1 denotes the identity, being performed on representations of the appropriate kind, and it is not difficult to verify the following relations

$$cr = 1 + t, \; c'q = 1 + t, \; rc = 2, \; qc' = 2, \; tr = r, \tag{2.13}$$

$$tc = c, \; tc' = c', \; qt = q, \; t^2 = 1.$$

For example, returning to the above splitting example, $c\psi = \varphi_1 + \varphi_2$ in (2.10), the relation $rc = 2$ tells us that $2\psi = r\varphi_1 + r\varphi_2$, namely $\psi = r\varphi_1 = r\varphi_2$.

As we have seen, a real representation $\psi \longleftrightarrow (G, U)$ can be extended to a complex representation $\varphi \longleftrightarrow (G, V)$ by the process of complexification, namely we set $V = U^{\mathbb{C}} = U + iU$ as in (2.3), and we set $\varphi = c\psi$. Conversely, we say that a complex representation $\varphi \longleftrightarrow (G, V)$ has a *real form* $\psi \longleftrightarrow (G, U)$ if $c\psi = \varphi$, that is, (G, V) is equivalent to $(G, U^{\mathbb{C}})$. In the same vein, a complex representation $\varphi \longleftrightarrow (G, V)$ has a *quaternionic form* $\eta \longleftrightarrow (G, W)$ if $c'\eta = \varphi$, that is (G, W) is equivalent to $(G, V^{\mathbb{H}})$ where G acts by $\mathbb{H}$-linear transformations on $V^{\mathbb{H}} = V + jV$.

Note that distinct real representations are mapped to distinct complex representations when they are complexified, and distinct quaternionic representations have distinct complex-reduction. Consequently, all the representation theory is actually contained in the realm of complex representations φ, regarding φ as *real* (resp. *quaternionic*) if it has a real (resp. quaternionic) form. The following general result is valid (at least) for compact groups G (for a proof, see e.g. [2]):

Lemma 2.1.2. *(a) Let φ be a complex representation of G. If φ is real or quaternionic, then φ is self-conjugate, that is, $t\varphi(= \bar{\varphi}) \simeq \varphi$. Conversely, if φ is irreducible and self-conjugate, then it is either real or quaternionic (but not both!).*

(b) φ is real (resp. quaternionic) if and only if (G, V) has a G-invariant non-singular symmetric (resp. skew-symmetric) bilinear form $V \times V \to \mathbb{C}$.

For any two representations of G on $\mathbb{K}$-vector spaces V_1 and V_2 respectively, set $Hom^G(V_1, V_2)$ to be the vector space of G-equivariant linear maps $F : V_1 \to V_2$. The following result is classical but of central importance.

Lemma 2.1.3 (Schur's lemma). *Let* $\varphi_i : G \to GL_{\mathbb{C}}(V_i)$, $i = 1, 2$, *be two irreducible representations of G on $\mathbb{C}$-vector spaces V_i. Then $Hom^G(V_1, V_2) \simeq \mathbb{C}$ if $\varphi_1 \simeq \varphi_2$, and $Hom^G(V_1, V_2) = 0$ otherwise.*

Proof. As $F : V_1 \to V_2$ is equivariant, it is easy to see that both $\ker(F)$ and $\text{Im}(F)$ are G-invariant subspaces and, from irreducibility, $F = 0$ or F is an isomorphism. In the latter case the two representations are equivalent, so let us assume $V_1 = V_2$. Since F has at least one eigenvalue and its corresponding eigenspace is G-invariant, it follows from irreducibility that F is a non-zero multiple of the identity. $\qquad\square$

2.1.3 The Infinitesimal Version of Lie Groups and Their Representations

Next we turn to the infinitesimal aspect of a connected Lie group G, namely the fundamental reduction to its *Lie algebra* $\mathcal{G}$, which is an $\mathbb{R}$-vector space, identified with the tangent space of G at the identity element e, endowed with a specific bilinear product

$$[\,,\,] : \mathcal{G} \times \mathcal{G} \to \mathcal{G}$$

called the *Lie bracket*, which is *skew-symmetric* and satisfies the *Jacobi identity*

$$[X, Y] = -[Y, X], \quad [[X, Y], Z] + [[Y, Z], X] + [[Z, X], Y] = 0.$$

The Lie bracket is, in fact, the "linearization" of the product in G, and this relation is more precisely understood via the so-called *exponential map* $\exp: \mathcal{G} \to G$, which provides the linkage between a connected group and its Lie algebra. It is a local diffeomorphism mapping a neighborhood of $X = 0$ onto a neighborhood of $e \in G$. This also explains why the Lie algebra is determined by any small neighborhood of $e \in G$, and consequently locally isomorphic groups have isomorphic Lie algebras. In particular, the Lie algebra of a disconnected Lie group G depends only on the identity component subgroup $G°$ of G.

Now, let us again focus attention on linear or matrix groups. The set $gl(V)$ of all $\mathbb{K}$-linear operators $T : V \to V$, resp. the set $gl_{\mathbb{K}}(n) = M_{\mathbb{K}}(n)$ of matrices, are vector spaces as well as associative algebras over $\mathbb{R}$ with regard to the usual product of operators, resp. matrices, and with the commutator product $[S, T] = ST - TS$ they are the Lie algebras of $GL(V)$ and $GL_{\mathbb{K}}(n)$, respectively. In these cases the exponential map is defined by the (usual) power series expansion

$$\exp(X) = e^X = I + X + \frac{1}{2}X^2 + \frac{1}{3!}X^3 + \ldots + \frac{1}{k!}X^k + \ldots. \qquad (2.14)$$

The inverse map is the logarithm

$$\log(I + Y) = \sum_{k=1}^{\infty} \frac{(-1)^{k+1}}{k} Y^k, \text{ for } |Y| < 1.$$

It follows, for example, that $\det(e^X) = e^{tr(X)}$, where $tr(X)$ is the trace of X.

For closed matrix groups $G \subset GL_{\mathbb{K}}(n)$, there is the following commutative diagram of groups, Lie algebras, and vertical exponential maps $\hat{e}$

$$\mathcal{G} \subset M_{\mathbb{K}}(n)$$
$$\hat{e} \downarrow \quad \downarrow \hat{e}$$
$$G \subset GL_{\mathbb{K}}(n).$$

Namely, the exponential map for G is the restriction of the matrix exponential map (2.14), and moreover, the Lie algebra $\mathcal{G}$ can be determined as

$$\mathcal{G} = \{X \in M_{\mathbb{K}}(n); \exp(tX) \in G \text{ for all } t\}. \tag{2.15}$$

As we shall mainly focus attention on orthogonal groups $SO(n) \subset O(n) \subset GL_{\mathbb{R}}(n)$ and unitary groups $SU(n) \subset U(n) \subset GL_{\mathbb{C}}(n)$, we first note that $O(n)$ has two connected components; the component different from $SO(n)$ consisting of the matrices with determinant -1. By (2.15), the Lie algebras of $SO(n)$ and $SU(n)$ consist of traceless skew-symmetric and skew-Hermitian matrices, respectively.

For compact connected groups G, such as $SO(n), SU(n), U(n)$, the exponential map is, in fact, surjective. But more generally, a connected Lie group G is still generated by its 1-parameter subgroups $\{\exp(tX), t \in \mathbb{R}\}$ and hence is determined by its Lie algebra, up to local isomorphisms. In fact, for a given basis $\{X_1, \ldots, X_r\}$ of $\mathcal{G}$, G is generated by the corresponding 1-parameter groups $\{\exp(tX_i)\}$. Therefore, the elements X_i are sometimes referred to as the infinitesimal generators of the group G, a terminology dating back to Sophus Lie in the nineteenth century.

Finally, an n-dimensional unitary representation φ of G yields by differentiation an associated representation φ_* of the Lie algebra $\mathcal{G}$, and there is the following commutative diagram

$$\mathcal{G} \xrightarrow{\varphi_*} \mathcal{U}(n)$$
$$\downarrow \hat{e} \quad \quad \downarrow \hat{e} \tag{2.16}$$
$$G \xrightarrow{\varphi} U(n)$$

where $\mathcal{U}(n)$ is the Lie algebra of $U(n)$ consisting of the skew-Hermitian matrices, and φ_* is a *Lie algebra homomorphism*, namely a linear map preserving the Lie bracket,

$$\varphi_*([X, Y]) = [\varphi_*(X), \varphi_*(Y)].$$

In effect, for a given set of infinitesimal generators $X_i \in \mathcal{G}$ for G, the representation φ is uniquely determined by the skew-Hermitian matrices $\tilde{X}_i = \varphi_*(X_i)$. In many cases, $\ker \varphi$ is a finite group, so that G is locally isomorphic with the image group $\tilde{G} = \varphi(G)$ in $U(n)$, wheras $\varphi_* : \mathcal{G} \to \tilde{\mathcal{G}} = \varphi_*(\mathcal{G})$ is an isomorphism between $\mathcal{G}$ and the Lie subalgebra $\tilde{\mathcal{G}}$ of $\mathcal{U}(n)$.

Conversely, a Lie algebra homomorphism of $\mathcal{G}$ is not always derived from a Lie group homomorphism φ of G, unless G is simply connected, such as the group $SU(n)$. In the general case, let us first replace G in (2.16) by the unique simply connected group $\bar{G}$ which is locally isomorphic to G, the so-called *universal covering group*. Then $G = \bar{G}/K$ for some discrete group K of $\bar{G}$, and the Lie algebra of $\bar{G}$ is still the same $\mathcal{G}$. Now, every homomorphism of $\mathcal{G}$ is derived from some φ for $\bar{G}$ as in the diagram (2.16), and this Lie algebra homomorphism will also exponentiate to a Lie group homomorphism of G if and only if $K \subset \ker \varphi$.

2.1.4 The Adjoint and the Coadjoint Representations

Let G be any Lie group with Lie algebra $\mathcal{G}$. Then G acts on itself by inner automorphisms $\mathrm{i}_g : G \to G$, $\mathrm{i}_g(h) = ghg^{-1}$, and differentiation of i_g at the identity $e \in G$ yields a homomorphism

$$Ad_G : G \to GL(\mathcal{G}), \ Ad_G(g) = (\mathrm{i}_g)_* : \mathcal{G} \to \mathcal{G}, \qquad (2.17)$$

where in fact $Ad_G(g)$ is a Lie algebra isomorphism. Ad_G is called the *adjoint* representation of G and the dual representation

$$Ad_G^T : G \to GL(\mathcal{G}^*)$$

is often called the *coadjoint* representation. These are real representations, and they are not necessarily equivalent, unless G is compact.

But in the latter case one can construct an equivalence, that is, a G-equivariant isomorphism $\Phi : \mathcal{G} \to \mathcal{G}^*$, by defining $\Phi(v) = v^*$ to be the linear functional $w \to \langle v, w \rangle$, where $\langle , \rangle$ denotes a G-invariant inner product on $\mathcal{G}$. The equivariance condition $\Phi(gv) = g\Phi(v)$ amounts to showing that $\Phi(Ad(g)v) = Ad^T(g)\Phi(v)$, and this is easily checked by applying both sides to a vector $w \in \mathcal{G}$.

For a compact connected Lie group, a characteristic property of the adjoint (or coadjoint) representation is that all orbits $\mathcal{O}$ are of type G/H, where H is a subgroup containing a maximal torus T. In fact, the principal orbits, filling an open dense subset of $\mathcal{G}$ (or $\mathcal{G}^*$), are all of type G/T. For a description of all these subgroups H of $G = SO(n)$, $SU(n)$, or $Sp(n)$, we refer to Table 1 in [68].

Finally, we recall that every coadjoint orbit $\mathcal{O}$ (and hence also an adjoint orbit when $\mathcal{O} = G/H$) naturally carries a G-invariant symplectic structure ω, as follows: For any $\mu \in \mathcal{O} \subset \mathcal{G}^*$ and $v, w \in \mathcal{G}$, via $(Ad_G^T)_*$ identify $v_\mu, w_\mu \in T_\mu \mathcal{O}$. Then,

$$\omega(v_\mu, w_\mu) = \mu([v, w]) \tag{2.18}$$

defines a G-invariant nondegenerate closed 2-form ω on $\mathcal{O}$, uniquely up to sign (cf. [48] for instance, for more details).

2.2 On the Lie Groups SU(2) and SO(3)

In this section we shall explore in greater detail the Lie groups $SU(2)$ and $SO(3)$, which have isomorphic Lie algebras; they are locally isomorphic groups since $SO(3) \simeq SU(2)/\{\pm Id\}$, with $SU(2)$ being the universal covering group of $SO(3)$.

2.2.1 Basic Definitions

Let $M_{\mathbb{K}}(n)$ be the $n \times n$-matrix algebra over $\mathbb{K} = \mathbb{R}$ or $\mathbb{C}$. First we focus attention on particular elements of $M_{\mathbb{C}}(2)$ and $M_{\mathbb{R}}(3)$ in order to exhibit the relationship between the groups $SU(2)$ and $SO(3)$, especially from the viewpoint of quantum mechanics. We recall that these groups are defined as follows:

$$SU(2) = \{g \in M_{\mathbb{C}}(2) \mid g^* g = g g^* = Id, \ \det g = 1\},$$

$$SO(3) = \{g \in M_{\mathbb{R}}(3) \mid g^T g = g g^T = Id, \ \det g = 1\}.$$

However, from the viewpoint of Lie group theory, the crucial fact is that $SO(3) \simeq SU(2)/\mathbb{Z}_2$, where $\mathbb{Z}_2 = \{\pm Id\}$ is the center of $SU(2)$.

Thus, we introduce the 3-vectors $\boldsymbol{\sigma} = (\sigma_1, \sigma_2, \sigma_3)$, $\mathbf{L} = (L_1, L_2, L_3)$ whose components are specific matrices

$$\sigma_1 = \begin{pmatrix} 0 & 1 \\ 1 & 0 \end{pmatrix}, \ \sigma_2 = \begin{pmatrix} 0 & -i \\ i & 0 \end{pmatrix}, \ \sigma_3 = \begin{pmatrix} 1 & 0 \\ 0 & -1 \end{pmatrix} \quad \text{(Pauli spin matrices)} \tag{2.19}$$

$$L_1 = \begin{pmatrix} 0 & 0 & 0 \\ 0 & 0 & -1 \\ 0 & 1 & 0 \end{pmatrix}, \ L_2 = \begin{pmatrix} 0 & 0 & 1 \\ 0 & 0 & 0 \\ -1 & 0 & 0 \end{pmatrix}, \ L_3 = \begin{pmatrix} 0 & -1 & 0 \\ 1 & 0 & 0 \\ 0 & 0 & 0 \end{pmatrix} \tag{2.20}$$

which provide natural bases for the Hermitian matrices and two Lie algebras,

$$\mathcal{H}(2) : I, \sigma_1, \sigma_2, \sigma_3$$

$$\mathcal{SU}(2) : i\sigma_1, i\sigma_2, i\sigma_3$$

$$\mathcal{SO}(3) : L_1, L_2, L_3.$$

We will also use the notation

$$\mathbf{n} \cdot \boldsymbol{\sigma} = \sum n_k \sigma_k, \quad \mathbf{n} \cdot \mathbf{L} = \sum n_k L_k$$

to denote any element of $\mathcal{H}(2)$ or $\mathcal{SO}(3)$, where the vector $\mathbf{n} \in \mathbb{R}^3$ is to be interpreted as pointing in the direction of the axis of rotation, in the case of $SO(3)$. To make this more precise, first observe that the commutation rules

$$\begin{cases} [\sigma_j, \sigma_k] = 2i\,\epsilon_{jkl}\sigma_l \\ [L_j, L_k] = \epsilon_{jkl}L_l \end{cases} \tag{2.21}$$

lead to a Lie algebra isomorphism

$$d\psi : \mathcal{SU}(2) \to \mathcal{SO}(3); \quad \begin{cases} \varepsilon_1 \frac{i}{2}\sigma_1 \to L_1 \\ \varepsilon_2 \frac{i}{2}\sigma_2 \to L_2 \\ \varepsilon_3 \frac{i}{2}\sigma_3 \to L_3 \end{cases} \tag{2.22}$$

for any choice of signs $\varepsilon_j = \pm 1$ with $\prod \varepsilon_j = -1$. Our standard choice will be $\varepsilon_j = -1$ for all j, and in terms of the exponential map, $A \to \exp(A) = e^A$, there is the geometrically suggestive notation

$$U(\mathbf{n}, \theta) = \exp(-\frac{i}{2}\theta(\mathbf{n} \cdot \boldsymbol{\sigma})), \quad R(\mathbf{n}, \theta) = \exp(\theta(\mathbf{n} \cdot \mathbf{L}))$$

which yields a "standard" homomorphism

$$\psi : SU(2) \to SO(3), \quad U(\mathbf{n}, \theta) \to R(\mathbf{n}, \theta). \tag{2.23}$$

We observe that, given a unit vector $\mathbf{n}$, $R(\mathbf{n}, \theta)$ is the rotation in euclidean 3-space through the angle θ (in the right-handed sense) around the axis directed along $\mathbf{n}$, namely

$$R(\mathbf{n}, \theta) : \mathbf{v} \to (\cos\theta)\mathbf{v} + (1 - \cos\theta)(\mathbf{n} \cdot \mathbf{v})\mathbf{n} + (\sin\theta)\mathbf{n} \times \mathbf{v}.$$

In view of this, $U(\mathbf{n}, \theta) \in SU(2)$ receives the quantum mechanical interpretation of a *spinor rotation* with respect to the axis determined by $\mathbf{n}$.

Since the homomorphism ψ is one-to-one for small θ (and **n** fixed), one can easily deduce the *adjoint formulas*

$$U(\mathbf{n} \cdot \boldsymbol{\sigma})U^{-1} = (R\mathbf{n}) \cdot \boldsymbol{\sigma}, \quad R(\mathbf{n} \cdot \mathbf{L})R^{-1} = (R\mathbf{n}) \cdot \mathbf{L}, \tag{2.24}$$

valid for any pair $U \in SU(2)$ and $R = \psi(U)$. In particular, this establishes a natural equivalence between the adjoint representation of $G = SU(2)$ or $SO(3)$, acting on its Lie algebra, and the standard representation of $SO(3)$, acting by rotations on euclidean 3-space $\mathbb{R}^3$.

Now, let $\mathbf{e}_i, i = 1, 2, 3$, be the usual orthonormal basis of $\mathbb{R}^3$, for which $\mathbf{e}_3 = (0, 0, 1)$ is identified with the north pole of the unit sphere

$$S^2 \subset \mathbb{R}^3 : x^2 + y^2 + z^2 = 1. \tag{2.25}$$

The classical way of expressing a rotation in terms of Euler angles (α, β, γ) is frequently used in quantum mechanics. We refer to [63], Chapter 13 or [74], Sect. 1.4. The idea is to express a rotation as a product of three "simple" rotations, namely rotations around two chosen coordinate axes. A widely used definition amounts to setting

$$R(\alpha, \beta, \gamma) = R(\mathbf{e}_3, \alpha) R(\mathbf{e}_2, \beta) R(\mathbf{e}_3, \gamma) = e^{\alpha L_3} e^{\beta L_2} e^{\gamma L_3} \tag{2.26}$$

$$U(\alpha, \beta, \gamma) = U(\mathbf{e}_3, \alpha) U(\mathbf{e}_2, \beta) U(\mathbf{e}_3, \gamma) = e^{-i \frac{\alpha}{2} \sigma_3} e^{-i \frac{\beta}{2} \sigma_2} e^{-i \frac{\gamma}{2} \sigma_3} \tag{2.27}$$

and then the above homomorphism (2.23) is expressed as

$$\psi : U(\alpha, \beta, \gamma) \to R(\alpha, \beta, \gamma).$$

We point out, however, that there is no canonical choice of homomorphism between $SU(2)$ and $SO(3)$. However, the various choices only differ by an automorphism of $SU(2)$ (or $SO(3)$). For example, complex conjugation in $SU(2)$ is the automorphism $\varsigma : U(\alpha, \beta, \gamma) \to U(-\alpha, \beta, -\gamma)$, which composed with ψ yields the homomorphism

$$\psi' = \psi \circ \varsigma : SU(2) \to SO(3), \ U(\alpha, \beta, \gamma) \to R(-\alpha, \beta, -\gamma). \tag{2.28}$$

In the following subsection we shall derive the two homomrphisms ψ and ψ' in a more geometric way, in terms of equivariant maps between spaces.

2.2.2 Hopf Map and Stereographic Projection

Now, $SU(2)$ acts by its standard (unitary) representation on the 3-sphere

$$S^3 \subset \mathbb{C}^2 : |z_1|^2 + |z_2|^2 = 1, \tag{2.29}$$

and the following diffeomorphism

$$\Psi : SU(2) \to S^3, \quad g = \begin{pmatrix} z_1 & -\bar{z}_2 \\ z_2 & \bar{z}_1 \end{pmatrix} \mapsto \Psi(g) = \begin{pmatrix} z_1 \\ z_2 \end{pmatrix} \sim (z_1, z_2) = \mathbf{z} \qquad (2.30)$$

which identifies $SU(2)$ with the 3-sphere, is equivariant when the group acts on itself by left translation (in Eq. (2.30) above, the symbol $\sim$ means that we identify a column vector with a line vector whenever the distinction is irrelevant, in order to save space). More precisely, equivariance under left action means that, for all g and $\Psi(g)$ as in (2.30) above, and for all

$$h = \begin{pmatrix} w_1 & -\bar{w}_2 \\ w_2 & \bar{w}_1 \end{pmatrix} \in SU(2) , \Psi(h) = \begin{pmatrix} w_1 \\ w_2 \end{pmatrix} \in S^3 \subset \mathbb{C}^2 ,$$

we have that

$$g\Psi(h) = \begin{pmatrix} z_1 & -\bar{z}_2 \\ z_2 & \bar{z}_1 \end{pmatrix} \begin{pmatrix} w_1 \\ w_2 \end{pmatrix} = \Psi \left(\begin{pmatrix} z_1 & -\bar{z}_2 \\ z_2 & \bar{z}_1 \end{pmatrix} \begin{pmatrix} w_1 & -\bar{w}_2 \\ w_2 & \bar{w}_1 \end{pmatrix} \right) = \Psi(gh) .$$

Let us recall the classical Hopf map

$$\pi : S^3 \to S^2 \simeq S^3 / U(1) = \mathbb{C}P^1, \quad \begin{cases} \mathbf{z} = (z_1, z_2) \to \mathbf{n} = (x, y, z) \\ x + iy = 2\bar{z}_1 z_2, \ z = |z_1|^2 - |z_2|^2 \end{cases}$$
$$(2.31)$$

which yields a fibration of the 3-sphere (2.29) over the 2-sphere (2.25). As indicated in (2.31), π also identifies S^2 with the orbit space of $U(1) = \{e^{i\theta}\}$ where $e^{i\theta}$ acts by scalar multiplication on vectors $\mathbf{z} \in \mathbb{C}^2$.

Now, $SO(3)$ acts by rotations on the 2-sphere, and associated with the map π is a distinguished homomorphism

$$\bar{\psi} : SU(2) \to SO(3) \qquad (2.32)$$

defined by the constraint that π is $\bar{\psi}$-equivariant, namely

$$\pi(g\mathbf{z}) = \bar{\psi}(g)\pi(\mathbf{z}), \ \text{for } g \in SU(2). \qquad (2.33)$$

In fact, $\bar{\psi}$ coincides with the homomorphism ψ in (2.23). Moreover, by writing $z_1 = x_1 + iy_1$, $z_2 = x_2 + iy_2$, a straightforward calculation of the homomorphism (2.32) gives the explicit expression

$$\psi(g) = \begin{pmatrix} (x_1^2 - x_2^2 - y_1^2 + y_2^2) & 2(x_1 y_1 - x_2 y_2) & x \\ -2(x_1 y_1 + x_2 y_2) & (x_1^2 + x_2^2 - y_1^2 - y_2^2) & y \\ -2(x_1 x_2 - y_1 y_2) & -2(x_1 y_2 + x_2 y_1) & z \end{pmatrix} \qquad (2.34)$$

where the third column is the vector $\mathbf{n}$ in (2.31). For historical reasons, the pair (z_1, z_2), subject to the condition $|z_1|^2 + |z_2|^2 = 1$, is also referred to as *Cayley-Klein* coordinates for $SU(2)$ and $SO(3)$.

Next, let us also recall the stereographic projection to the complex plane $\mathbb{C}$

$$\pi' : S^2 - \{\mathbf{e}_3\} \to \mathbb{C}, (x, y, z) \to \xi = \frac{x + iy}{1 - z} \tag{2.35}$$

and the action of $SU(2)$ on $\mathbb{C}$ by fractional linear (or Möbius) transformations

$$\xi \to \frac{z_1 \xi - \bar{z}_2}{z_2 \xi + \bar{z}_1}. \tag{2.36}$$

Associated with the map (2.35) is a homomorphism $SU(2) \to SO(3)$ which makes (the inverse of) the 1-1 correspondence π' in (2.35) equivariant, and by straightforward calculations the homomorphism is found to be ψ' in (2.28).

The identity $Id \in SU(2)$ corresponds via Ψ in (2.30) to the basic vector $\mathbf{z} = (1, 0)$, which by π is mapped to the north pole $\mathbf{e}_3$ of S^2, and the corresponding isotropy groups $H = U(1) \simeq SO(2)$ in $G = SU(2)$ and $SO(3)$ are related by

$$SU(2) \supset \left\{ e^{-i\frac{\theta}{2}\sigma_3} \right\} = U(1) \to^{\psi} SO(2) = \left\{ e^{\theta L_3} \right\} \subset SO(3). \tag{2.37}$$

This also realizes the 2-sphere as a homogeneous space in two ways,

$$\begin{array}{l} SU(2) = S^3 \searrow^{\pi} \\ \quad \downarrow \psi \qquad\quad S^2 = G/H = SU(2)/U(1) = SO(3)/SO(2) \\ SO(3) \nearrow_{\bar{\pi}} \end{array} \tag{2.38}$$

where $\bar{\pi} : SO(3) \to S^2$ projects a matrix to its third column $\mathbf{n} = (x, y, z)$.

$SO(3)$ acts by orthogonal transformations on $\mathbb{R}^3$; this is the standard representation ρ_3. The adjoint representation $Ad_{SO(3)}$ is the action by conjugation, $Ad(g)S = gSg^{-1}$, on skew symmetric matrices $S \in \mathcal{SO}(3)$. We set up the following 1-1 correspondence between $\mathcal{SO}(3)$ and $\mathbb{R}^3$

$$S = \begin{pmatrix} 0 & z & y \\ -z & 0 & x \\ -y & -x & 0 \end{pmatrix} \longleftrightarrow \begin{pmatrix} x \\ y \\ z \end{pmatrix}$$

which is, in fact, $SO(3)$-equivariant, and this proves that ρ_3 identifies with the adjoint, hence also the coadjoint representation of $SO(3)$.

The orbits are therefore all concentric spheres $S^2(r)$ of radius $r \geq 0$, which for $r > 0$ are of type $SO(3)/SO(2)$, as in (2.38).

It is often convenient to have expressions for functions and various other structures defined on S^2 written in local coordinates. Here we shall mostly write them in local spherical polar coordinates (θ, φ) where, for simplicity, we identify S^2 with the unit sphere in $\mathbb{R}^3$ according to (2.25), so that (θ, φ) are defined by

$$x = \sin \varphi \cos \theta, \quad y = \sin \varphi \sin \theta, \quad z = \cos \varphi \tag{2.39}$$

where φ is the colatitude, θ is the longitude and the origin of the polar coordinate system is the north pole $\mathbf{e}_3 = \mathbf{n}_0 = (0, 0, 1) \in S^2(1) \subset \mathbb{R}^3$.

Note that, since the radius of any 2-sphere in $\mathbb{R}^3$ is an $SO(3)$-invariant quantity, we are free to rescale all spheres of radius $r > 0$ to the unit sphere. Or equivalently, by using angular coordinates (θ, φ) we can simply forget about radii.

In particular, the G-invariant symplectic (or area) form, cf. (2.18), is (locally) expressed in terms of spherical polar coordinates as

$$\omega = \sin \varphi d\varphi \wedge d\theta . \tag{2.40}$$

Of course, instead of polar coordinates we could use complex coordinates on S^2, but a relation between these two local coordinate systems is straightforwardly obtained from (2.39) and (2.35).

2.2.3   Prelude to the Irreducible Unitary Representations of SU(2)

The irreducible (unitary) representations of $SU(2)$ are finite dimensional and typically denoted by $[j]$ in the physics literature, but we shall also use the notation

$$\varphi_j = [j], \quad j = 0, 1/2, 1, 3/2, \ldots : \dim_{\mathbb{C}} \varphi_j = 2j + 1, \tag{2.41}$$

These are $SO(3)$-representations only when j is an integer l, in which case they have a real form

$$\psi_l, \quad l = 0, 1, 2, 3, \ldots, \dim_{\mathbb{R}} \psi_l = 2l + 1, \tag{2.42}$$

that is, $\varphi_l = [\psi_l]^{\mathbb{C}}$ is the complexification of ψ_l. However, if j has half-integral value, φ_j has a quaternionic form, so, in both cases φ_j is actually self-dual, that is, $\bar{\varphi}_j \simeq \varphi_j$ for all j (another issue is the precise relation of a given basis with its dual and the precise form of this equivalence).

It is a basic fact about compact connected Lie groups G that its maximal tori T constitute a single conjugacy class (T), and each element $g \in G$ can be conjugated into a fixed torus T, say $hgh^{-1} = t \in T$ for some $h \in G$. In particular, for the matrix groups $SU(n)$, $U(n)$ this is the diagonalization of matrices, when T is chosen to be the diagonal matrices with entries $e^{i\theta_k}$. As a consequence of this, a representation of G is uniquely determined by its restriction to the torus T.

In the case of $G = SU(2)$, the diagonal group $U(1)$ in (2.37) is our chosen maximal torus T. By Schur's lemma (cf. Lemma 2.1.3), a unitary representation φ of $SU(2)$, when restricted to $U(1)$, splits into 1-dimensional representations and clearly each one is determined by a homomorphism $U(1) \to \mathbb{C}^*$ of type

$$diag(e^{i\theta}, e^{-i\theta}) \to e^{qi\theta}, q \in \mathbb{Z}$$

where the "functional" $q\theta$ is called the *weight* of the above $U(1)$-representation. Thus, there are altogether $(n + 1) = \dim \varphi$ weights $q_i\theta$, and the totality of weights

$$\Omega(\varphi) = \{q_0\theta, q_1\theta, q_2\theta, \ldots, q_n\theta\} \tag{2.43}$$

completely determines φ and is referred to as the *weight system* of the $SU(2)$-representation (with respect to $U(1)$).

Remark 2.2.1. The collection (2.43) must be regarded as a *multiset*, namely the elements are counted with multiplicity, since the $n + 1$ weights are not necessarily distinct but the total multiplicity equals $n + 1$. For example, as multisets we write identities of the following type:

$$\{a, a, b, c, b, a, c, b, d, e\} = 3\{a, b\} + 2\{c\} + \{d, e\}.$$

Now, a well-known way to build an $(n + 1)$-dimensional representation of $SU(2)$ explicitly from its standard 2-dimensional representation is by mapping $\mathbb{C}^2$ to the space of complex homogeneous polynomials of degree n in two variables:

$$\mathbf{z} = (z_1, z_2) \in \mathbb{C}^2 \to \mathbb{C}^{n+1} \simeq {}^h P_{\mathbb{C}}^n(z_1, z_2) = Span_{\mathbb{C}}\{z_1^{n-k} z_2^k\}_{0 \le k \le n}.$$

Then, as $SU(2)$ acts on $\mathbb{C}^2$ via its standard representation, this induces an action of $SU(2)$ on ${}^h P_{\mathbb{C}}^n(z_1, z_2)$ and the $SU(2)$ representations obtained in this way turn out to be irreducible. A choice of orthonormal basis for ${}^h P_{\mathbb{C}}^n(z_1, z_2)$ is the ordered set $\{\mathbf{v}(n, k)\}$, where

$$\mathbf{v}(n, k) = \sqrt{\binom{n}{k}} z_1^{n-k} z_2^k, k = 0, 1, \cdots, n. \tag{2.44}$$

However, such a basis is often written in terms of $j = n/2$ and $m = j - k$.

As will be made clearer further below, Eq. (2.44) defines the basis $\{\mathbf{v}(n, k)\}$, for each $n = 2j$, only up to an overall phase factor. In fact, accounting for a scaling freedom for the inner product on ${}^h P_{\mathbb{C}}^n(z_1, z_2)$ implies that two such bases given by (2.44) can be identified, for each j, if they differ from each other by an overall non-zero complex number (cf. Schur's Lemma 2.1.3), and the basis introduced by Bargmann [7] differs from (2.44) above by the $\sqrt{n!}$ factor.

Finally, note that one can map these concrete representations of $SU(2)$ on $^hP_{\mathbb{C}}^n(z_1, z_2)$ to concrete representations on the space of n-degree holomorphic polynomials on S^2, $\mathcal{H}ol^n(S^2)$, by composing with the projective maps

$$\mathbb{C}^2 - \{0\} \to \mathbb{C}P^1 \simeq S^2 \,, \ (z_1, z_2) \mapsto (1, \xi = z_2/z_1), \ \text{or } (\zeta = z_1/z_2, 1).$$

Composing (2.44) with the projective map $(z_1, z_2) \mapsto \zeta$ yields the basis used by Berezin [12], while the map $(z_1, z_2) \mapsto \xi$ and Bargmann's normalization convention yields the basis used in the book of Vilenkin and Klimyk [75], where explicit expressions for irreducible representations of $SU(2)$ are presented.

In the next chapter, the irreducible representations of $SU(2)$ shall be studied in great detail in terms of infinitesimal techniques involving weights, operators and Lie algebras, which is the approach commonly used in quantum mechanics.

Chapter 3
Quantum Spin Systems and Their Operator Algebras

This chapter presents the basic mathematical framework for quantum mechanics of spin systems. Much of the material can be found in texts in representation theory (some found within the list of references at the beginning of Chap. 2) and quantum theory of angular momentum (e.g. [13, 14, 16, 23, 46, 63, 65], some of these being textbooks in quantum mechanics which can also be used by the reader not too familiar with the subject as a whole). Our emphasis here is to provide a self-contained presentation of quantum spin systems where, in particular, the combinatorial role of Clebsch-Gordan coefficients and various kinds of Wigner symbols is elucidated, leading to the $SO(3)$-invariant decomposition of the operator product which, strangely enough, we have not found explicitly done anywhere.

3.1 Basic Definitions of Quantum Spin Systems

In line with the standard formulation of quantum affine mechanical systems, we define quantum spin systems as follows:

Definition 3.1.1. A *spin-j quantum mechanical system*, or *spin-j system*, is a complex Hilbert space $\mathcal{H}_j \simeq \mathbb{C}^{n+1}$ together with an irreducible unitary representation

$$\varphi_j : SU(2) \to G \subset U(\mathcal{H}_j) \simeq U(n+1), \quad n = 2j \in \mathbb{N}, \tag{3.1}$$

where G denotes the image of $SU(2)$ and hence is isomorphic to $SU(2)$ or $SO(3)$ according to whether j, called the *spin number*, is half-integral or integral.

Remark 3.1.2. Unless otherwise stated (as in Appendix 10.8), throughout this monograph we shall always assume the Hermitian inner product of a Hilbert space is skew-linear (conjugate linear) in the first variable.

© Springer International Publishing Switzerland 2014
P. de M. Rios, E. Straume, *Symbol Correspondences for Spin Systems*,
DOI 10.1007/978-3-319-08198-4_3

A vector in $\mathcal{H}_j$ is also called a j-*spinor*. For our description of bases, operators and matrix reprentations we will use familiar terminology and notation from quantum mechanics. The representation (3.1) is normally described at the infinitesimal level by Hermitian operators J_1, J_2, J_3 satisfying the standard commutation relations for angular momentum, namely

$$[J_a, J_b] = i\epsilon_{abc} J_c \qquad (3.2)$$

together with the basic relation

$$\mathbf{J}^2 = J_1^2 + J_2^2 + J_3^2 = j(j+1)I . \qquad (3.3)$$

Indeed, by (3.2) the operator sum $\mathbf{J}^2$ commutes with each J_i, so by irreducibility it must be a multiple of the identity. All this is equivalent to saying that the group G with the Lie algebra

$$\mathcal{G} = lin_{\mathbb{R}}\{iJ_1, iJ_2, iJ_3\} \simeq \mathcal{SU}(2) \qquad (3.4)$$

of skew-Hermitian operators acts irreducibly on $\mathcal{H}_j$.

In analogy with $\boldsymbol{\sigma}$ and $\mathbf{L}$ in (2.19), (2.20), the vector of operators

$$\mathbf{J} = (J_1, J_2, J_3) \qquad (3.5)$$

is referred to as the *total angular momentum (or spin)* operator of the quantum system, and its components satisfy the standard commutation relations (3.2) for angular momentum. The commutation relations are also symbolically expressed as $\mathbf{J} \times \mathbf{J} = i\mathbf{J}$, and the square $\mathbf{J}^2$ is the operator sum (3.3).

Remark 3.1.3. In the physics literature, one usually finds Planck's constant $\hbar$, resp. $\hbar^2$, explicitly multiplying the r.h.s. of Eq. (3.2), resp. Eq. (3.3), which also guarantees that $\mathbf{J}$ has the dimensions of angular momentum. However, since this factor can be removed by an appropriate scaling of $\mathbf{J}$, we will omit it throughout almost the whole book.

Note that the infinitesimal generators $-iJ_k, k = 1, 2, 3$, of the operator group G satisfy the same commutation relations as the operators L_k in (2.21), that is, the correspondence $L_k \to -iJ_k, 1 \le k \le 3$, is a Lie algebra isomorphism, and for a unit vector $\mathbf{n}$ in euclidean 3-space we shall refer to the operator

$$J_\mathbf{n} = \mathbf{n} \cdot \mathbf{J} = \sum n_i J_i \qquad (3.6)$$

as the *angular momentum (or spin) in the direction of* $\mathbf{n}$. The corresponding homomorphism in (3.1) is

$$\varphi_j : e^{\frac{1}{2}i\theta(\mathbf{n} \cdot \boldsymbol{\sigma})} \to e^{i\theta(\mathbf{n} \cdot \mathbf{J})}. \qquad (3.7)$$

3.1.1 Standard Basis and Standard Matrix Representations

The above operators on $\mathcal{H}_j$ will be represented by matrices with respect to a suitable choice of orthonormal basis, unique up to a common phase factor. We'll adopt the generally adopted convention, expressed in the partial definition below:

Definition 3.1.4 (partial). A basis that diagonalizes the operator $J_{\mathbf{n}_0} = J_3$, for $\mathbf{n}_0$ identified with the positive z axis, shall be referred to as a *standard basis* for $\mathcal{H}_j$.[1]

This basis will be fully characterized below in terms of the action of angular momentum operators. Starting with the simplest case, for a spin-$\frac{1}{2}$ quantum system the angular momentum (3.5) is defined to be the following vector of operators

$$\mathbf{J} = \frac{1}{2}\,\sigma = \frac{1}{2}(\sigma_1, \sigma_2, \sigma_3), \tag{3.8}$$

namely the Pauli matrices with the factor $1/2$. Then $G = SU(2)$ in (3.1), and $\varphi_{1/2}$ is the identity. In particular, the cartesian basis $\{\mathbf{e}_1, \mathbf{e}_2\}$ of $\mathcal{H}_{1/2} = \mathbb{C}^2$ diagonalizes J_3 with eigenvalues $\pm 1/2$, and this is a standard basis, see below.

Next, for a spin-1 quantum system the angular momentum is defined to be

$$\mathbf{J} = i\mathbf{L}$$

and a standard basis is typically chosen to be

$$-\frac{1}{\sqrt{2}}(\mathbf{e}_1 + i\,\mathbf{e}_2), \mathbf{e}_3, \frac{1}{\sqrt{2}}(\mathbf{e}_1 - i\,\mathbf{e}_2). \tag{3.9}$$

(Note that the first and last elements in this standard basis are *complex* linear combinations of $\mathbf{e}_1$ and $\mathbf{e}_2$, cf. Remark 3.1.8, below).

In general, for a spin-j system one would like to "measure" the angular momentum of $\mathbf{J}$ in a chosen direction $\mathbf{n}$. The eigenvalues of $J_\mathbf{n}$ are sometimes referred to as *magnetic quantum numbers*.

By choosing the direction $\mathbf{n}$ to be the positive z axis $\mathbf{n}_0$, cf. Definition 3.1.4, one obtains a standard orthonormal basis for $\mathcal{H}_j$ ordered by the magnetic quantum numbers m of $J_{\mathbf{n}_0} = J_3$, denoted as follows:

$$\mathbf{u}(j, m) = |jm\rangle, \quad m = j, j - 1, \ldots, -j + 1, -j, \tag{3.10}$$

consisting of eigenvectors of J_3 whose eigenvalues constitute the string of numbers m as indicated in (3.10), where Dirac's "ket" notation for the vectors is displayed. Thus, with the above ordering of the basis, J_3 has the matrix representation

---

[1]When $\mathbf{n}_0$ is seen as a point on the unit sphere $S^2 \subset \mathbb{R}^3$, it is also called the "*north pole*".

$$J_3 = \begin{bmatrix} j & 0 & 0 & 0 & 0 \\ 0 & j-1 & 0 & 0 & 0 \\ \vdots & \vdots & \vdots & \vdots & \vdots \\ 0 & 0 & 0 & -j+1 & 0 \\ 0 & 0 & 0 & 0 & -j \end{bmatrix} \tag{3.11}$$

So far, however, the vectors in the above basis (3.10) are only determined modulo an individual phase factor.

Thus, in terms of weights we consider the circle subgroup $\{e^{i\theta\sigma_3}\}$ of $SU(2)$ (cf. (2.37)), consisting of the spinor rotations around the z-axis, acting on $\mathcal{H}_j$ with the vectors (3.10) as weight vectors and $2m\theta$ as the associated weight, namely

$$\varphi_j(e^{i\theta\sigma_3}) = e^{i2\theta J_3} : \mathbf{u}(j,m) \to e^{2mi\theta}\mathbf{u}(j,m). \tag{3.12}$$

Therefore, by definition, the weight system of φ_j is the set

$$\Omega(\varphi_j) = \{2j\theta, 2(j-1)\theta, \ldots, -2j\theta\}. \tag{3.13}$$

Hence, in order to fix a phase convention for a standard basis (3.10), which also fixes our *standard* matrix representation of the operators $J_i, i = 1, 2, 3$, let us first invoke the structure of the algebra (3.4), expressed by the commutation rules (3.2). To this end one introduces the mutually adjoint pair of operators

$$J_+ = J_1 + iJ_2, \quad J_- = J_1 - iJ_2 \tag{3.14}$$

called the *raising* and *lowering* operators, respectively, whose commutation rules

$$[J_+, J_-] = 2J_3, \quad [J_3, J_\pm] = \pm J_\pm, \tag{3.15}$$

yield the following identity between nonnegative Hermitian operators

$$J_-J_+ = J^2 - J_3(J_3 + I). \tag{3.16}$$

The relations (3.15) also imply

$$J_+\mathbf{u}(j,m) = \alpha_{j,m}\mathbf{u}(j,m+1), \quad J_-\mathbf{u}(j,m) = \beta_{j,m}\mathbf{u}(j,m-1), \tag{3.17}$$

for some constants $\alpha_{j,m}, \beta_{j,m}$ which are non-zero, except that $\alpha_{jj} = \beta_{j,-j} = 0$ (since there is no eigenvector outside the range (3.10)).

Then we finally arrive at our complete definition:

Definition 3.1.5 (complete). A *standard basis* $\{\mathbf{u}(j,m)\}$ of $\mathcal{H}_j$, ordered as in (3.10), is defined by choosing the first (and highest weight) unit vector $\mathbf{u}(j,j)$ and inductively fixing the phase of $\mathbf{u}(j,m-1)$ so that $\beta_{j,m}$ in (3.17) is always nonnegative.

Consequently, a standard basis is unique up to one common phase factor $e^{i\omega}$. The above commutation rules yield the formulae

$$\alpha_{j,m} = \sqrt{(j-m)(j+m+1)}, \quad \beta_{j,m} = \sqrt{(j+m)(j-m+1)}. \quad (3.18)$$

With respect to a standard basis the mutually adjoint matrices representing $J_\pm$ have nonnegative entries, so they are the transpose of each other, with all non-zero entries on a subdiagonal, as illustrated (where $n = 2j$):

$$J_- = J_+^T = \begin{bmatrix} 0 & 0 & 0 & 0 & & 0 & 0 \\ \sqrt{n \cdot 1} & 0 & 0 & 0 & & 0 & 0 \\ 0 & \sqrt{(n-1) \cdot 2} & 0 & 0 & & 0 & 0 \\ \vdots & \vdots & \vdots & \vdots & & \vdots & \vdots \\ 0 & 0 & 0 & \sqrt{2 \cdot (n-1)} & & 0 & 0 \\ 0 & 0 & 0 & 0 & & \sqrt{1 \cdot n} & 0 \end{bmatrix}. \quad (3.19)$$

From this, one calculates the Hermitian matrices

$$J_1 = \frac{1}{2}(J_+ + J_-), \quad J_2 = \frac{1}{2i}(J_+ - J_-),$$

and in the initial case $j = 1/2$ the identity (3.8) is recovered.

Finally, consider also the induced action of $SU(2)$ on the dual space

$$\mathcal{H}_j^* = Hom(\mathcal{H}_j, \mathbb{C}) \simeq \mathcal{H}_j, \quad (3.20)$$

namely the dual representation $\bar{\varphi}_j$. This is isomorphic to φ_j, so we may regard the two spaces in (3.20) as being the same underlying Hilbert space. Then the two cases are distinguished by their actions, namely $g \in SU(2)$ acts by its complex conjugate $\bar{g}$ in the dual case. The resulting effect on infinitesimal generators is that J_2 is invariant, whereas J_1 and J_3 are multiplied by -1, and thus the raising and lowering operators in the dual case are

$$\check{J}_+ = -J_-, \quad \check{J}_- = -J_+.$$

Consequently, in view of (3.17) a standard basis of $(\varphi_j, \mathcal{H}_j)$ is not a standard basis of the dual representation $(\check{\varphi}_j, \check{\mathcal{H}}_j) \equiv (\bar{\varphi}_j, \mathcal{H}_j^*)$. However, the standard basis with vectors in the opposite order and with alternating sign changes is a dual standard basis. Our choice of sign convention is specified as follows:

Definition 3.1.6. The dual standard basis, dual to $\{\mathbf{u}(j, m)\}$, is the ordered collection of vectors

$$\check{\mathbf{u}}(j, m) = (-1)^{j+m} \mathbf{u}(j, -m), \quad -j \le m \le j. \quad (3.21)$$

Remark 3.1.7. Observe that the standard duality (3.21) is not "involutive" when j is half-integral, since applying the dual construction twice amounts to multiplying the original vectors $\mathbf{u}(j, m)$ by $(-1)^{2j}$.

The unitary operators $\varphi_j(g)$ on $\mathcal{H}_j$ are represented by well-defined unitary matrices $D^j(g)$. Using the notation (2.27) for elements $g \in SU(2)$, we shall denote the corresponding unitary operators $\varphi_j(g)$ on $\mathcal{H}_j$ by $\hat{D}^j(g)$ or $\hat{D}^j(\alpha, \beta, \gamma)$, namely we have the homomorphism

$$\varphi_j : U(\alpha, \beta, \gamma) \rightarrow \hat{D}^j(\alpha, \beta, \gamma) = e^{-i\alpha J_3} e^{-i\beta J_2} e^{-i\gamma J_3}. \tag{3.22}$$

The associated matrix of $\hat{D}^j$ with respect to a standard basis (3.10) is the matrix $D^j = \left(D^j_{m_1, m_2} \right)$ whose entries are the following functions on $SU(2)$:

$$D^j_{m_1, m_2}(g) = \left\langle \mathbf{u}(j, m_1), \hat{D}^j(g) \mathbf{u}(j, m_2) \right\rangle, \tag{3.23}$$

also called the *Wigner D-functions*, cf. [74], Chap. 4.

Remark 3.1.8. The reader should be aware that even when j is an integer, so that the $SU(2)$-representation is effectively a representation of $SO(3)$, the standard representation is a complex representation. Thus, for instance, the standard representation of $SU(2)$ for $j = 1$ consists of complex 3×3 matrices. Namely, by conjugation with the unitary transition matrix from the basis $\mathbf{e}_i, i = 1, 2, 3$, to the basis (3.9), the real matrix group with Lie algebra generated by (2.20) becomes a complex matrix group.

3.2   The Tensor Product and the Space of Operators

For a given spin-j quantum mechanical system $\mathcal{H}_j$, let us identify the Hilbert space with the complex $(n + 1)$-space $\mathbb{C}^{n+1}, n = 2j$, by the correspondence

$$\mathbf{e}_k = |j, j - k + 1\rangle, \quad k = 1, 2, \ldots, n + 1, \tag{3.24}$$

which identifies a given standard basis (3.10) of $\mathcal{H}_j$ with the usual standard basis of $\mathbb{C}^{n+1}$, namely the column matrices

$$\mathbf{e}_1 = (1, 0, \ldots, 0)^T, \mathbf{e}_2 = (0, 1, 0, \ldots, 0)^T, \text{ etc.}$$

For two systems $\mathcal{H}_{j_1}, \mathcal{H}_{j_2}$, the space of linear operators $Hom(\mathcal{H}_{j_2}, \mathcal{H}_{j_1})$ identifies with the full matrix space $M_{\mathbb{C}}(n_1 + 1, n_2 + 1)$, linearly spanned by the one-element matrices

$$\mathcal{E}_{k,l} = \mathbf{e}_k \mathbf{e}_l^T \text{ (matrix product)}, \quad (\mathcal{E}_{kl})_{pq} = \delta_{kp}\delta_{lq}, \tag{3.25}$$

and there is the linear isometry

$$\mathbb{C}^{n_1+1} \otimes \mathbb{C}^{n_2+1} \rightarrow M_{\mathbb{C}}(n_1 + 1, n_2 + 1), \quad \mathbf{e}_k \otimes \mathbf{e}_l \rightarrow \mathcal{E}_{kl} \tag{3.26}$$

where the matrix space has the (Hilbert-Schmidt) Hermitian inner product

$$\langle P, Q \rangle = trace(P^*Q) = \text{Re} \langle P, Q \rangle + i \,\text{Im} \langle P, Q \rangle \tag{3.27}$$

and $P^* = \overline{P}^T$ is the adjoint of P. The real part in (3.27) is a euclidean metric for the matrix space viewed as a real vector space.

We are primarily interested in the case $n_1 = n_2 = n$, in which case the matrix space, denoted by $M_{\mathbb{C}}(n + 1)$, is also an algebra and has the orthogonal decomposition

$$M_{\mathbb{C}}(n + 1) = \mathcal{A}Sym(n + 1) \oplus Sym(n + 1) \tag{3.28}$$

into skew-symmetric and symmetric matrices. Moreover, as a real vector space there is the real orthogonal decomposition (w.r.t. the real part in (3.27))

$$M_{\mathbb{C}}(n + 1) = \mathcal{U}(n + 1) \otimes_{\mathbb{R}} \mathbb{C} = \mathcal{U}(n + 1) \oplus \mathcal{H}(n + 1), \tag{3.29}$$

where $\mathcal{U}(n+1)$ is the Lie algebra of $U(n+1)$ consisting of skew-Hermitian matrices and $\mathcal{H}(n + 1) = i\,\mathcal{U}(n + 1)$ is the space of Hermitian matrices.

3.2.1 SU(2)-Invariant Decomposition of the Tensor Product

Let μ_{n+1} be the standard representation of $U(n + 1)$ on $\mathbb{C}^{n+1}$, and let $\check{\mu}_{n+1}$ be its dual with $g \in U(n + 1)$ acting by the complex conjugate matrix $\bar{g}$ on $\mathbb{C}^{n+1}$. Consider the tensor product representations $\mu_{n_1+1} \otimes \mu_{n_2+1}$ and $\mu_{n_1+1} \otimes \check{\mu}_{n_2+1}$ of $U(n_1 + 1) \times U(n_2 + 1)$ acting on $\mathbb{C}^{n_1+1} \otimes \mathbb{C}^{n_2+1}$. The matrix model of these representations follows from the isometry (3.26), when the group acts on matrices $P \in M_{\mathbb{C}}(n_1 + 1, n_2 + 1)$ by matrix multiplication, as follows:

$$(i) \quad \mu_{n_1+1} \otimes \mu_{n_2+1} : (g, h)P \rightarrow gPh^T, \tag{3.30}$$

$$(ii) \quad \mu_{n_1+1} \otimes \check{\mu}_{n_2+1} : (g, h)P \rightarrow gPh^{-1}.$$

Composing with irreducible representations $\varphi_{j_i} : SU(2) \rightarrow U(n_i + 1)$, yields the following tensor product representations of $SU(2)$ and its action on matrices:

$$(i) \; \varphi_{j_1} \otimes \varphi_{j_2} : (g, P) \to \varphi_{j_1}(g) P \varphi_{j_2}(g)^T, \qquad (3.31)$$

$$(ii) \; \varphi_{j_1} \otimes \bar{\varphi}_{j_2} : (g, P) \to \varphi_{j_1}(g) P \varphi_{j_2}(g)^{-1}.$$

However, since the $SU(2)$-representations $\bar{\varphi}_j$ and φ_j are equivalent for any j, so are the two tensor products and their equivariant matrix models (3.31). Combining (3.24), (3.26), and Definition 3.1.6, we are led to the following:

Definition 3.2.1. For the two matrix models (3.31) of the tensor product $\mathbb{C}^{n_1+1} \otimes \mathbb{C}^{n_2+1}$, the *uncoupled standard basis* is the following collection of one-element matrices (cf. (3.21)):

$$\text{model (i):} \; |j_1 m_1 j_2 m_2\rangle = \mathbf{u}(j_1, m_1) \otimes \mathbf{u}(j_2, m_2) = \mathcal{E}_{j_1-m_1+1, j_2-m_2+1}, \quad (3.32)$$

$$\text{model (ii):} \; |j_1 m_1 j_2 m_2\rangle = \mathbf{u}(j_1, m_1) \otimes \breve{\mathbf{u}}(j_2, m_2) = (-1)^{j_2+m_2} \mathcal{E}_{j_1-m_1+1, j_2+m_2+1},$$

where $-j_i \le m_i \le j_i$, and all $j_i - m_i$ are integers.

At the infinitesimal level the angular momentum operators $J_k, k = 1, 2, 3$, of $SU(2)$ act on matrices P in the two models by

$$(i) \; J_k \cdot P = J_k^{(j_1)} P + P(J_k^{(j_2)})^T, \quad (ii) \; J_k \cdot P = J_k^{(j_1)} P - P J_k^{(j_2)}, \qquad (3.33)$$

where the matrix $J_k^{(j)} \in M_{\mathbb{C}}(2j + 1)$ represents J_k acting on $\mathcal{H}_j = \mathbb{C}^{2j+1}$. In particular, for $1 \le k \le n_1 + 1, 1 \le l \le n_2 + 1$,

$$(i) \; J_3 \cdot \mathcal{E}_{kl} = (j_1 + j_2 + 2 - k - l)\mathcal{E}_{kl}, \quad (ii) \; J_3 \cdot \mathcal{E}_{kl} = (j_1 - j_2 + l - k)\mathcal{E}_{kl} \qquad (3.34)$$

and this tells us that the uncoupled basis of $M_{\mathbb{C}}(n_1 + 1, n_2 + 1)$ diagonalizes J_3, with the eigenvalues as shown in (3.34).

Remark 3.2.2. Thus, in model (i) the eigenspace of *quantum magnetic number* $m = m_1 + m_2$ consists of the "anti-subdiagonal" matrices spanned by matrices $\mathcal{E}_{kl}$ with $k + l = (j_1 + j_2 + 2 - m)$, whereas in model (ii) the eigenspace is the "subdiagonal" spanned by the matrices $\mathcal{E}_{kl}$ with $l - k = j_2 - j_1 + m$, so that in model (ii) $\mathcal{E}_{kl}$ is the actual m-th subdiagonal when $j_1 = j_2$.

Next, let us decompose the tensor product (3.30) into irreducible summands, by first calculating the weight system of the tensor product and then determining its decomposition, using the fact that a representation ϕ is uniquely determined by its weight system $\Omega(\phi)$. For convenience, let us formally define

$$\{a_1, a_2, \ldots, a_m\} \otimes \{b_1, b_2, \ldots, b_n\} = \{(a_i + b_j); 1 \le i \le m, 1 \le j \le n\}$$

to be the "tensor product" of two multisets (cf. Remark 2.2.1). Observe that the vector $|j_1 m_1 j_2 m_2\rangle$ in (3.32) is a weight vector of weight $2\theta(m_1 + m_2)$ in the tensor product (3.31). Setting $\lambda = j_1 + j_2 - |j_1 - j_2| + 1$ and omitting (for convenience) the factor 2θ of the weights, then, by writing a union of multisets additively (again, see Remark 2.2.1), we have by (3.13)

$$\Omega(\varphi_{j_1} \otimes \varphi_{j_2}) = \{j_1, j_1 - 1, \ldots, -j_1\} \otimes \{j_2, j_2 - 1, \ldots, -j_2\}$$

$$= \sum_{k=1}^{\lambda} k \{\pm(j_1 + j_2 - k + 1)\} = \sum_{j=|j_1-j_2|}^{j_1+j_2} \{j, j - 1, \ldots, -j\}$$

$$\tag{3.35}$$

$$= \sum_{j=|j_1-j_2|}^{j_1+j_2} \Omega(\varphi_j).$$

Remark 3.2.3. A neat way to describe the range of j in the sum (3.35) is to state $\delta(j_1, j_2, j) = 1$. This is the "triangle inequality" condition, involving three nonnegative integral or half-integral numbers. Namely,

$$\delta(j_1, j_2, j_3) = 1 \iff$$

$$(i) \ |j_1 - j_2| \le j_3 \le j_1 + j_2 \ \text{ and } \ (ii) \ j_1 + j_2 + j_3 \in \mathbb{Z}, \tag{3.36}$$

and $\delta(j_1, j_2, j_3) = 0$ otherwise. We must note that the condition is symmetric, that is, independent of the order of the numbers.

It follows from (3.35) that

$$\varphi_{j_1} \otimes \varphi_{j_2} = \sum_{\delta(j_1, j_2, j)=1} \varphi_j \tag{3.37}$$

and each of the summands φ_j in (3.37) has its own standard basis, denoted by

$$|(j_1 j_2)jm\rangle, m = j, j - 1, \ldots, -j + 1, -j \tag{3.38}$$

in the literature, and *this basis is unique up to a phase factor for each j.* By (3.32) these vectors are identified with specific matrices in $M_{\mathbb{C}}(n_1 + 1, n_2 + 1)$, and as pointed out this can be done naturally in two different ways depending on the choice of matrix model. In any case, there is the orthogonal decomposition

$$M_{\mathbb{C}}(n_1 + 1, n_2 + 1) = \sum_{j=|j_1-j_2|}^{j_1+j_2} M_{\mathbb{C}}(\varphi_j)$$

where $M_{\mathbb{C}}(\varphi_j)$ has the standard orthonormal basis (3.38), for each j. The totality of these vectors (or matrices) constitute the *coupled standard basis* of the tensor product (or matrix space).

Clebsch-Gordan Coefficients

To describe the connection between the coupled and uncoupled basis the following definition is crucial.

Definition 3.2.4. The *Clebsch-Gordan* coefficients are, by definition, the entries of the unitary transition matrix relating the uncoupled and coupled standard basis, namely the inner products

$$C_{m_1,m_2,m}^{j_1,j_2,j} = \langle (j_1 j_2) j m | j_1 m_1 j_2 m_2 \rangle \tag{3.39}$$

which are the coefficients in the expansion

$$|j_1 m_1 j_2 m_2 \rangle = \sum_{j=|j_1-j_2|}^{j_1+j_2} \sum_{m=-j}^{j} C_{m_1,m_2,m}^{j_1,j_2,j} |(j_1 j_2) j m \rangle . \tag{3.40}$$

Remark 3.2.5. Clebsch-Gordan coefficients, also called Wigner coefficients, have been extensively studied in the physics literature; we refer to [13, 24, 74] for surveys of their properties. First, they satisfy the following non-vanishing conditions:

$$C_{m_1,m_2,m}^{j_1,j_2,j} \neq 0 \Longrightarrow \begin{cases} m = m_1 + m_2 \\ \delta(j_1, j_2, j) = 1 \end{cases} . \tag{3.41}$$

Also, they are uniquely determined once a phase connvention for the coupled basis is chosen, and on the other hand, such a convention follows by choosing the phase of some of the coefficients. We shall follow the generally accepted convention (cf. e.g. [13, 24, 74])

$$C_{j_1,j-j_1,j}^{j_1,j_2,j} > 0 \quad \text{whenever } \delta(j_1, j_2, j) = 1 , \tag{3.42}$$

which, in fact, also implies that *all the coefficients are real*. Then, it follows from their definition that they satisfy the following orthogonality equations:

$$\sum_{m_1,m_2} C_{m_1,m_2,m}^{j_1,j_2,j} C_{m_1,m_2,m'}^{j_1,j_2,j'} = \delta_{j,j'}\delta_{m,m'} , \quad \sum_{j,m} C_{m_1,m_2,m}^{j_1,j_2,j} C_{m_1',m_2',m}^{j_1,j_2,j} = \delta_{m_1,m_1'}\delta_{m_2,m_2'} . \tag{3.43}$$

Consequently, the unitary transition matrix in the above definition is orthogonal, so the inversion of (3.40) is the formula

$$|(j_1 j_2) j m \rangle = \sum_{(m_1+m_2=m)} C_{m_1,m_2,m}^{j_1,j_2,j} |j_1 m_1 j_2 m_2 \rangle . \tag{3.44}$$

In particular, when $m = j_1 + j_2$ there is only one term in the expansion (3.40), so $C_{j_1,j_2,j_1+j_2}^{j_1,j_2,j_1+j_2} = 1$. Moreover, when $j_1 = 0$ or $j_2 = 0$, there is no reason to distinguish between φ_j, $\varphi_0 \otimes \varphi_j$, and $\varphi_j \otimes \varphi_0$, and so $C_{m,0,m}^{j,0,j} = C_{0,m,m}^{0,j,j} = 1$.

On the other hand, there is a connection between Clebsch-Gordan coefficients and the Wigner D-functions defined by (3.23), given by the following *coupling rule*:

Proposition 3.2.6. *For a fixed $g \in SU(2)$,*

$$D_{\mu_1,m_1}^{j_1} D_{\mu_2,m_2}^{j_2} = \sum_j C_{\mu_1,\mu_2,\mu_1+\mu_2}^{j_1,j_2,j} C_{m_1,m_2,m_1+m_2}^{j_1,j_2,j} D_{\mu_1+\mu_2,m_1+m_2}^{j} \qquad (3.45)$$

whose inversion formula reads

$$D_{\mu m}^{j} = \sum_{\mu_1} \sum_{m_1} C_{m_1,m_2,m}^{j_1,j_2,j} C_{\mu_1,\mu_2,\mu}^{j_1,j_2,j} D_{\mu_1,m_1}^{j_1} D_{\mu_2,m_2}^{j_2} . \qquad (3.46)$$

We refer to Appendix 10.1 for a proof of Proposition 3.2.6.

Remark 3.2.7. The above proposition shows that the Wigner D-functions are essentially determined by the Clebsch-Gordan coefficients and vice-versa.

Thus, starting from the trivially available four functions $\{D_{kl}^{1/2}\}$ one can use the formula (3.46) to determine successively the functions $D_{\mu m}^{j}$ for all j.

Conversely, starting from the expression (3.19) and formulas (3.14)–(3.15) and (3.22)–(3.23) which produce formulas for the $D_{\mu m}^{j}$, one can use formula (3.46) to compute all Clebsch-Gordan coefficients explicitly. This is the way these coefficients were first explicitly computed, by Wigner in 1927 [88, 89].

In fact, iterating the recursive Eq. (3.68) below, properly generalized to $|(j_1 j_2)jm\rangle$, gives another way to obtain explicit expressions for all Clebsch-Gordan coefficients, as explored by Racah. Thus, there are various equivalent explicit expressions for the Clebsch-Gordan coefficients, see for instance [13, 74]. Here we list a rather symmetric one, first obtained by van der Waerden in 1932:

$$C_{m_1,m_2,m_3}^{j_1,j_2,j_3} = \delta_{m_3,m_1+m_2} \sqrt{2j+1} \, \Delta(j_1, j_2, j_3) \, S_{m_1,m_2,m_3}^{j_1, j_2, j_3} \qquad (3.47)$$

$$\cdot \sum_z \frac{(-1)^z}{z!(j_1+j_2-j_3-z)!(j_1-m_1-z)!(j_2+m_2-z)!(j_3-j_2+m_1+z)!(j_3-j_1-m_2+z)!}$$

where by definition

$$\Delta(j_1, j_2, j_3) = \sqrt{\frac{(j_1 + j_2 - j_3)!(j_3 + j_1 - j_2)!(j_2 + j_3 - j_1)!}{(j_1 + j_2 + j_3 + 1)!}} , \qquad (3.48)$$

$$S_{m_1,m_2,m_3}^{j_1, j_2, j_3} = \sqrt{(j_1 + m_1)!(j_1 - m_1)!(j_2 + m_2)!(j_2 - m_2)!(j_3 + m_3)!(j_3 - m_3)!} \qquad (3.49)$$

Remark 3.2.8. We recall that the Clebsch-Gordan coefficients are non-zero and satisfy Eq. (3.47) above only if the conditions (3.41) are satisfied. Also, in the sum $\sum_z$ of formula (3.47), the summation index z is assumed to take all integral values for which all factorial arguments are nonnegative, with the usual convention $0! = 1$. In the sequel we shall also encounter similar summations, and the same convention on the summation index is tacitly assumed unless otherwise stated.

By inspection of this and other equivalent formulae, one obtains the symmetry properties for the Clebsch-Gordan coefficients. Here we list some of these:

$$C_{m_1,m_2,m_3}^{j_1,j_2,j_3} = (-1)^{j_1+j_2-j_3} C_{-m_1,-m_2,-m_3}^{j_1,j_2,j_3} = (-1)^{j_1+j_2-j_3} C_{m_2,m_1,m_3}^{j_2,j_1,j_3}, \tag{3.50}$$

$$C_{m_1,m_2,m_3}^{j_1,j_2,j_3} = (-1)^{j_2+m_2} \sqrt{\frac{2j_3+1}{2j_1+1}} C_{-m_3,m_2,-m_1}^{j_3,j_2,j_1}, \tag{3.51}$$

$$C_{m_1,m_2,m_3}^{j_1,j_2,j_3} = (-1)^{j_1-m_1} \sqrt{\frac{2j_3+1}{2j_2+1}} C_{m_1,-m_3,-m_2}^{j_1,j_3,j_2}. \tag{3.52}$$

3.3   SO(3)-Invariant Decomposition of the Operator Algebra

We shall further investigate the special case $j_1 = j_2 = j$ and resume the terminology from the previous section. In particular, the matrix algebra $M_{\mathbb{C}}(n+1)$ represents the space of linear operators on $\mathcal{H}_j = \mathbb{C}^{n+1}$, on which the unitary group $U(n+1)$ acts by two different (inner) tensor product representations

$$(i)\ \ \mu_{n+1} \otimes \mu_{n+1} \simeq \Lambda^2 \mu_{n+1} + S^2 \mu_{n+1} : (g, P) \to gPg^T , \tag{3.53}$$

$$(ii)\ \ \mu_{n+1} \otimes \breve{\mu}_{n+1} \simeq Ad_{U(n+1)}^{\mathbb{C}} =_{\mathbb{R}} 2Ad_{U(n+1)} : (g, P) \to gPg^{-1}.$$

The splitting in the two cases corresponds to the $U(n+1)$-invariant decompositions (3.28) and (3.29), respectively. In case (ii) the splitting is over $\mathbb{R}$ and $U(n+1)$ acts by its (real) adjoint representation $Ad_{U(n+1)}$ on both $\mathcal{U}(n+1)$ and $\mathcal{H}(n+1)$.

Composition of the above representations with the irreducible representation $\varphi_j : SU(2) \to U(n+1)$ yields the following two equivalent representations

$$(i)\ \ \varphi_j \otimes \varphi_j \simeq (\Lambda^2 \mu_{n+1} + S^2 \mu_{n+1}) \circ \varphi_j : (g, P) \to \varphi_j(g) P \varphi_j(g)^T, \tag{3.54}$$

$$(ii)\ \ \varphi_j \otimes \bar{\varphi}_j \simeq Ad_{U(n+1)}^{\mathbb{C}} \circ \varphi_j : (g, P) \to \varphi_j(g) P \varphi_j(g)^{-1}.$$

According to (3.37) this representation splits into an integral string of irreducibles

$$\varphi_j \otimes_{\mathbb{C}} \varphi_j = \varphi_0 + \varphi_1 + \ldots + \varphi_n, \ n = 2j. \tag{3.55}$$

Let us denote the corresponding decomposition of the matrix space as

$$M_{\mathbb{C}}(n+1) = \sum_{l=0}^{n} M_{\mathbb{C}}(\varphi_l), \tag{3.56}$$

where the summands consist of either symmetric or skew-symmetric matrices, depending on the parity of l and according to the splitting

$$S^2 \mu_{n+1} | SU(2) = \varphi_n + \varphi_{n-2} + \varphi_{n-4} + \ldots,$$
$$\Lambda^2 \mu_{n+1} | SU(2) = \varphi_{n-1} + \varphi_{n-3} + \varphi_{n-5} + \ldots.$$

The above tensor product (3.55) is, in fact, a representation of $SO(3) = SU(2)/\mathbb{Z}_2$ and hence it has a real form. Such a real form can be embedded in $M_{\mathbb{C}}(n+1)$ in different ways; for example as the space of real matrices

$$M_{\mathbb{R}}(n+1) = \sum_{l=0}^{n} M_{\mathbb{R}}(\psi_l), \tag{3.57}$$

where the irreducible summands consist of either symmetric or skew-symmetric matrices, depending on the parity of l as in (3.56), cf. (2.41) and (2.42).

Definition 3.3.1. In order to agree with the standard framework in quantum mechanics (see Remark 3.3.13), we henceforth stick to the **matrix model (ii)** in (3.54) and therefore $SO(3)$ acts via the adjoint action of $U(n+1)$ on $M_{\mathbb{C}}(n+1)$.

Thus, the above representation of $SO(3)$ on the matrix space (3.56) splits into real invariant subspaces

$$\mathcal{U}(n+1) = \sum_{l=0}^{n} \mathcal{U}(\psi_l) \,, \ \mathcal{H}(n+1) = \sum_{l=0}^{n} \mathcal{H}(\psi_l) \,, \tag{3.58}$$

$$\mathcal{H}(\psi_l) = \mathcal{H}(n+1) \cap M_{\mathbb{C}}(\varphi_l) \,, \mathcal{U}(\psi_l) = \mathcal{U}(n+1) \cap M_{\mathbb{C}}(\varphi_l).$$

At the infinitesimal level the angular momentum operators J_k, represented as matrices in $\mathcal{H}(n+1)$, act on $M_{\mathbb{C}}(n+1)$ via the commutator product

$$J_k \cdot P = ad_{J_k}(P) = [J_k, P] = J_k P - P J_k, \ k = 1, 2, 3$$
$$J_3 \cdot \mathcal{E}_{kl} = [J_3, \mathcal{E}_{kl}] = (l - k)\mathcal{E}_{kl}, \ 1 \leq k, l \leq n+1$$

For example, the summand $M_{\mathbb{C}}(\varphi_0)$ (resp. $M_{\mathbb{C}}(\varphi_1)$) is linearly spanned by the identity matrix I (resp. the matrices J_k).

Let us introduce the J_3-eigenspace decompositions

$$M_{\mathbb{C}}(n+1) = \sum_{m=-n}^{n} \Delta_{\mathbb{C}}(m), \quad M_{\mathbb{R}}(n+1) = \sum_{m=-n}^{n} \Delta_{\mathbb{R}}(m) \tag{3.59}$$

where $\Delta(m)$ consists of the m-subdiagonal matrices, spanned by the one-element matrices $\mathcal{E}_{k,l}$ with $l - k = m$ (cf. (3.25) and Remark 3.2.2). Clearly

$$\dim_{\Bbbk} \Delta_{\Bbbk}(m) = n + 1 - |m|.$$

In particular, the zero weight space $\Delta(0)$ consists of the main diagonal matrices, and $\Delta(n)$ (resp. $\Delta(-n)$) is spanned by the one-element matrix $\mathcal{E}_{1,n+1}$ (resp. $\mathcal{E}_{n+1,1}$) with its non-zero entry positioned at the upper right (resp. lower left) corner. It is sometimes convenient to denote an m-subdiagonal matrix $P = (P_{ij})$ with m-subdiagonal entries x_i as a coordinate vector

$$P = (x_1, x_2, \ldots, x_k)_m, k = n + 1 - |m|, \tag{3.60}$$

3.3.1 The Irreducible Summands of the Operator Algebra

We shall further investigate how the irreducible summands $M_{\mathbb{C}}(\varphi_l)$ (resp. $\mathcal{U}(\psi_l)$) and $\mathcal{H}(\psi_l)$) are embedded in the operator (or matrix) algebra $M_{\mathbb{C}}(n+1)$. It suffices to consider the Hermitian operators since $\mathcal{U}(\psi_l)) = i\,\mathcal{H}(\psi_l)$ and

$$M_{\mathbb{C}}(\varphi_l) = \mathcal{H}(\psi_l) + i\mathcal{H}(\psi_l).$$

To this end, consider the subspace

$$\mathcal{H}(n+1)_l \subset \mathcal{H}(n+1)$$

of Hermitian operators formally expressible as real homogeneous polynomials P, Q. of degree l in the non-commuting "variables" J_k. As generators of the Lie algebra $\mathcal{SO}(3)$ the operators $L_k = -iJ_k$ act as derivations on polynomials,

$$ad_{L_k}(PQ) = ad_{L_k}(P)Q + P\,ad_{L_k}(Q); \quad ad_{L_a}(J_b) = \varepsilon_{abc}J_c$$

and this action preserves the degree of a polynomial, leaving $\mathcal{H}(n+1)_l$ invariant.

The non-commutativity of the operators J_k can be handled by considering ordered 3-partitions of l,

$$\pi = (l_1, l_2, l_3), \ l_i \geq 0, \sum l_i = l.$$

For each such partition π there is a symmetric polynomial expression in the symbols J_i

$$P_\pi = J_1^{l_1} J_2^{l_2} J_3^{l_3} + \ldots + \qquad (3.61)$$

obtained from the leading monomial by symmetrization, as indicated in (3.61), that is, P_π is the sum of all monomials with the same total degree l_k for each J_k. For example, associated with $\pi = (1, 2, 0)$ is the polynomial $J_1 J_2^2 + J_2 J_1 J_2 + J_2^2 J_1$. The monomials in (3.61) are actually equal as operators, modulo a polynomial of lower degree, due to the basic commutation relation (3.2). Also, for $l = 0$ the only operator of type (3.61) is $P = Id$.

Let us denote by δ_l the number of partitions π of the above kind. We claim that for $l \leq n$ the operators P_π constitute a basis for $\mathcal{H}(n+1)_l$, and consequently

$$\dim \mathcal{H}(n+1)_l = \delta_l = \frac{1}{2}(l+1)(l+2) .$$

The crucial reason is that the (Casimir) operator $J^2 = J_1^2 + J_2^2 + J_3^2$, which by (3.3) acts as a multiple of the identity, yields by multiplication an $\mathcal{SO}(3)$-invariant imbedding

$$\mathcal{H}(n+1)_{l-2} \to J^2 \mathcal{H}(n+1)_{l-2} \subset \mathcal{H}(n+1)_l$$

and there is a complementary and invariant subspace $\mathcal{V}_l \subset \mathcal{H}(n+1)_l$, namely $\mathcal{V}_l = \mathcal{H}(\psi_l)$, of dimension

$$\dim \mathcal{V}_l = \delta_l - \delta_{l-2} = 2l + 1.$$

Thus, we have

$$\sum_{l=0}^{n} \dim \mathcal{V}_l = \sum_{l=0}^{n} (2l + 1) = (n+1)^2 = \dim \mathcal{H}(n+1)$$

and in fact,

$$\mathcal{H}(n+1) = \sum_{l=0}^{n} \mathcal{V}_l = \sum_{l=0}^{n} \mathcal{H}(\psi_l)$$

is the splitting (3.58). In particular, each monomial of degree $n+1$ is actually an operator expressible as a linear combination of monomials of degree $\leq n$.

Example 3.3.2. For fixed $n > 1$ the operators

$$A_1 = J_2 J_3 + J_3 J_2, \; A_2 = J_3 J_1 + J_1 J_3, \; A_3 = J_1 J_2 + J_2 J_1,$$
$$B_1 = J_2^2 - J_3^2, \; B_2 = J_3^2 - J_1^2, \; B_3 = J_1^2 - J_2^2 \quad (B_1 + B_2 + B_3 = 0)$$

in the 6-dimensional space $\mathcal{H}(n+1)_2$ span a 5-dimensional $SO(3)$-invariant subspace V_2, whose orthogonal complement is a trivial summand, namely the line spanned by the operator J^2.

3.3.2 The Coupled Standard Basis of the Operator Algebra

The last example does not provide any clue to the calculation of standard bases for the irreducible summands $M_{\mathbb{C}}(\varphi_l)$ of $M_{\mathbb{C}}(n+1)$. What we seek is a collection of matrices

$$\mathbf{e}^j (l, m) = |(jj)lm\rangle, \; -l \le m \le l, \; 0 \le l \le n = 2j \tag{3.62}$$

such that for each l the matrices $\mathbf{e}^j (l, m)$ constitute a standard basis for $M_{\mathbb{C}}(\varphi_l)$. Moreover, with the phase convention (3.42), namely the positivity condition

$$C_{j,l-j,l}^{j,\,j,l} > 0 \text{ for each } l , \tag{3.63}$$

the basis will be uniquely determined.

Remark 3.3.3. In what follows, we sometimes drop the superscript j for the coupled basis vectors $\mathbf{e}^j (l, m)$, when there is no ambiguity about the total spin j.

Now, we shall construct the coupled standard basis (3.62), expressed in terms of the matrices $J_\pm$, see (3.19). Consider the repeated product of J_+ with itself

$$(J_+)^l \in \Delta_{\mathbb{R}}(l) \subset M_{\mathbb{R}}(n+1),$$

noting that the product vanishes for $l = n + 1$, and by definition, $(J_\pm)^0 = I$. For a fixed n, the norms of the above matrices yield a sequence of positive integers depending on n,

$$\mu_l^n = \| (J_+)^l \| = \sqrt{trace((J_-)^l (J_+)^l)}, \; 0 \le l \le n, \tag{3.64}$$

where

$$\mu_0^n = \sqrt{n+1}, \; \mu_n^n = n!, \tag{3.65}$$

and moreover, for $l \ge 0$ there is the general formula

$$(\mu_l^n)^2 = \frac{(l!)^2}{(2l+1)!}(n-l+1)(n-l+2)\cdots n(n+1)(n+2)\cdots(n+l+1). \quad (3.66)$$

As a function of n this is, in fact, a polynomial of degree $2l+1$.

Definition 3.3.4. For fixed $l \geq 0$, define recursively a string of real matrices of norm 1,

$$\mathbf{e}^j(l,m) \in \Delta_{\mathbb{R}}(m), \quad -l \leq m \leq l \quad (3.67)$$

by setting

$$\mathbf{e}^j(l,l) = \frac{(-1)^l}{\mu_l^n}(J_+)^l, \quad \mathbf{e}^j(l,m-1) = \frac{1}{\beta_{l,m}}\left[J_-, \mathbf{e}^j(l,m)\right], \quad (3.68)$$

where $\beta_{l,m}$ is the number defined in (3.18).

Proposition 3.3.5. *The above family of real matrices* $\mathbf{e}^j(l,m)$ *constitute a coupled standard orthonormal basis*

$$|(jj)lm\rangle = \mathbf{e}^j(l,m); \quad 0 \leq l \leq n = 2j, \ -l \leq m \leq l,$$

for $M_{\mathbb{C}}(n+1)$ *in agreement with the phase convention (3.63)*

$$C^{j,j,l}_{j,l-j,l} = (-1)^l \langle \mathbf{e}^j(l,l), \mathcal{E}_{1,l+1}\rangle = (-1)^l \mathbf{e}^j(l,l)_{1,l+1} > 0. \quad (3.69)$$

For fixed l, the family of vectors $\mathbf{e}^j(l,m)$ *is a standard orthonormal basis for the irreducible tensor summand* $M_{\mathbb{C}}(\varphi_l)$*, and as matrices satisfy the relations*

$$\mathbf{e}^j(l,-m) = (-1)^m \mathbf{e}^j(l,m)^T, \quad -l \leq m \leq l, \quad (3.70)$$

and in particular,

$$\mathbf{e}^j(l,-l) = \frac{1}{\mu_l^n}(J_-)^l.$$

Proof. For a fixed l, if we have a unit vector $\mathbf{u}(l,l) \in M_{\mathbb{C}}(\varphi_l)$ with maximal J_3-eigenvalue $m = l$ and thus belonging to the l-subdiagonal $\Delta_{\mathbb{C}}(l)$, then successive application of the lowering operator J_- and normalization of the vectors will generate an orthonormal basis $\{\mathbf{u}(l,m)\}$ for $M_{\mathbb{C}}(\varphi_l)$ which, by Definition 3.1.5, is a standard basis. The "higher level" vectors $\mathbf{u}(l+1,l),\ldots,\mathbf{u}(n,l)$ span a hyperplane of $\Delta_{\mathbb{C}}(l)$, so knowledge of these vectors would uniquely determine $\mathbf{u}(l,l)$ up to a choice of phase.

We claim that one can take $\mathbf{u}(l,m)$ to be the above matrix $\mathbf{e}(l,m)$ for all (l,m). The point is that

$$\Delta_{\mathbb{R}}(l) = lin\{\mathbf{e}(l,l), \mathbf{e}(l+1,l), \ldots, \mathbf{e}(n,l)\} \tag{3.71}$$

where the listed vectors, indeed, constitute an orthonormal basis. To see this, we may assume inductively that the "higher level" vectors $\mathbf{e}(l+k,l)$ in (3.71) are already known to be perpendicular, and then it remains to check that $\mathbf{e}(l,l)$ is perpendicular to all $\mathbf{e}(l+k,l)$, $k > 0$. However, their inner product with $\mathbf{e}(l,l)$ is (modulo a factor $\neq 0$)

$$trace((J_+^l)^T ad(J_-)^k(J_+^{l+k})) = trace(J_-^l \left[J_-, \ldots \left[J_-, J_+^{l+k} \right] \ldots \right]) = 0,$$

by successive usage of the rule $trace(XY) = trace(YX)$.

Observe that the real matrix $(J_+)^l$ is l-subdiagonal and with positive entries. In particular, its inner product with $\mathcal{E}_{1,l+1}$ is positive. On the other hand, by model (ii) in (3.32)

$$\mathbf{e}(j, l - j) = (-1)^l \mathcal{E}_{1,l+1}$$

and thus the factor $(-1)^l$ in (3.68) is needed because of the sign convention (3.69).

Finally, the identity (3.70) can be seen from symmetry considerations using that J_- is the transpose of J_+, but with due regard to the sign convention.   □

Now, we shall obtain an explicit general expression for the coupled basis vectors $\mathbf{e}(l,m)$. But first, let us look at the unnormalized matrices $E(l,m)$, namely

$$E(l,m) = (-1)^l \mu_{l,m}^n \mathbf{e}(l,m), \quad \mu_{l,m}^n = \|E(l,m)\| . \tag{3.72}$$

They are constructed recursively, as follows. For $0 \le l \le n, -l \le m \le l$, define

$$E(l,l) = J_+^l, \quad E(0,0) = Id, \tag{3.73}$$

$$E(l,m-1) = [J_-, E(l,m)] = ad_{J_-}(E(l,m))$$

and hence there is the general formula

$$E(l,m) = (ad_{J_-})^{l-m}(J_+^l) = \sum_{k=0}^{l-m} (-1)^k \binom{l-m}{k} J_-^{l-m-k} J_+^l J_-^k. \tag{3.74}$$

It remains to determine the norm $\mu_{l,m}^n$ of $E(l,m)$, see (3.72). First, we remark there is the following identity

$$[J_+, E(l,m)] = \alpha_{l,m}^2 E(l, m + 1) , \text{ cf. (3.18)} \tag{3.75}$$

and next, for fixed l define positive numbers $p_{l,m}$ recursively by

$$p_{l,0} = 1, \quad p_{l,m+1} = \alpha_{l,m}^2 p_{l,m}, \quad 0 < m \leq l.$$

Then, the matrices $E(l, m)$ and $E(l, -m)$ are related via transposition by

$$E(l, -m) = (-1)^m p_{l,m} E(l, m)^T, \quad m \geq 0. \tag{3.76}$$

Finally, it follows from (3.73), (3.64), and (3.18) that

$$\mu_{l,l}^n = \mu_l^n, \quad \mu_{l,m-1}^n = \mu_{l,m}^n \beta_{l,m}$$

and consequently

$$\mu_{l,m}^n = \frac{l!}{\sqrt{2l+1}} \sqrt{\frac{(n+l+1)!}{(n-l)!}} \sqrt{\frac{(l-m)!}{(l+m)!}}, \quad 0 \leq m \leq l, \tag{3.77}$$

where we used Eq. (3.66) for μ_l^n. Thus, from (3.74) we obtain:

Theorem 3.3.6. *The coupled standard basis vectors of $M_{\mathbb{C}}(n+1)$ are given by*

$$\mathbf{e}^j(l, -m) = (-1)^m \mathbf{e}^j(l, m)^T \text{ for } -l \leq m \leq 0, \text{ where for } 0 \leq m \leq l,$$

$$\mathbf{e}^j(l, m) = \frac{(-1)^l}{\mu_{l,m}^n} \sum_{k=0}^{l-m} (-1)^k \binom{l-m}{k} J_-^{l-m-k} J_+^l J_-^k, \tag{3.78}$$

with $\mu_{l,m}^n$ given by (3.77) and $J_\pm \in M_{\mathbb{R}}(n+1)$ given by (3.19).

Remark 3.3.7. In accordance with Eq. (3.68) in Definition 3.3.4, the matrices $\mathbf{e}^j(l, m)$ given explicitly by (3.78) above satisfy

$$[J_+, \mathbf{e}^j(l, m)] = \alpha_{l,m} \mathbf{e}^j(l, m), \quad [J_-, \mathbf{e}^j(l, m)] = \beta_{l,m} \mathbf{e}^j(l, m), \tag{3.79}$$

with $\alpha_{l,m}$ and $\beta_{l,m}$ as in (3.18) and $J_\pm$ given by (3.19). However, one can also verify the following relations:

$$[J_3, \mathbf{e}^j(l, m)] = m \mathbf{e}^j(l, m), \quad \sum_{k=1}^3 [J_k, [J_k, \mathbf{e}^j(l, m)]] = l(l+1)\mathbf{e}^j(l, m), \tag{3.80}$$

where $J_1 = (J_+ + J_-)/2$, $J_2 = (J_+ - J_-)/2i$ and J_3 is given by (3.11).

Now, for the sake of clarifying another interpretation for this basis, let us rewrite the usual adjoint action of an operator A on an operator B as the action of a "superoperator" $\mathbb{A}$ associated to A as follows:

$$\mathbb{A} \cdot B = [A, B] = ad_A(B).$$

Then, we can rewrite (3.80) as

$$\mathbb{J}_3 \cdot e^j(l,m) = me^j(l,m), \quad \mathbb{J}^2 \cdot e^j(l,m) = l(l+1)e^j(l,m), \tag{3.81}$$

so that the $(2l+1)$-dimensional subspace $\mathcal{V}_l \subset M_{\mathbb{C}}(n+1)$ is the eigenspace of the "superoperator" $\mathbb{J}^2 = \mathbb{J}_1^2 + \mathbb{J}_2^2 + \mathbb{J}_3^2$ of eigenvalue $l(l+1)$.

In the physics literature, one usually interprets the tensor product $\varphi_{j_1} \otimes \varphi_{j_2}$ as a "sum", so that the coupled invariant spaces are eigenspaces of "addition of angular momenta". But in our context, we take the tensor product $\varphi_j \otimes \bar{\varphi}_j$ and therefore "$\mathbb{J}$" is better interpreted as the "difference of angular momenta" (we thank Robert Littlejohn for this point).

Illustrations

We shall illustrate the general formula (3.78) above by generating the matrices $e^j(l,m)$ in the lower dimensional cases, showing explicitly that each $e^j(l,m)$ is in fact an m-subdiagonal matrix.

Example 3.3.8. $j = 1/2$:

$$e(1,1) = \begin{pmatrix} 0 & -1 \\ 0 & 0 \end{pmatrix} \quad e(1,0) = \frac{1}{\sqrt{2}}\begin{pmatrix} 1 & 0 \\ 0 & -1 \end{pmatrix} \quad e(1,-1) = \begin{pmatrix} 0 & 0 \\ 1 & 0 \end{pmatrix}$$

Example 3.3.9. $j = 1$:

$$e(2,2) = \begin{pmatrix} 0 & 0 & 1 \\ 0 & 0 & 0 \\ 0 & 0 & 0 \end{pmatrix} \quad e(2,1) = \frac{1}{\sqrt{2}}\begin{pmatrix} 0 & -1 & 0 \\ 0 & 0 & 1 \\ 0 & 0 & 0 \end{pmatrix} \quad e(2,0) = \frac{1}{\sqrt{6}}\begin{pmatrix} 1 & 0 & 0 \\ 0 & -2 & 0 \\ 0 & 0 & 1 \end{pmatrix}$$

$$e(1,1) = \frac{-1}{\sqrt{2}}\begin{pmatrix} 0 & 1 & 0 \\ 0 & 0 & 1 \\ 0 & 0 & 0 \end{pmatrix} \quad e(1,0) = \frac{1}{\sqrt{2}}\begin{pmatrix} 1 & 0 & 0 \\ 0 & 0 & 0 \\ 0 & 0 & -1 \end{pmatrix}$$

Example 3.3.10. $j = 3/2$:

$$e(3,3) = \begin{pmatrix} 0 & 0 & 0 & -1 \\ 0 & 0 & 0 & 0 \\ 0 & 0 & 0 & 0 \\ 0 & 0 & 0 & 0 \end{pmatrix} \quad e(3,2) = \frac{1}{\sqrt{2}}\begin{pmatrix} 0 & 0 & 1 & 0 \\ 0 & 0 & 0 & -1 \\ 0 & 0 & 0 & 0 \\ 0 & 0 & 0 & 0 \end{pmatrix}$$

$$e(3,1) = \frac{1}{\sqrt{5}}\begin{pmatrix} 0 & -1 & 0 & 0 \\ 0 & 0 & \sqrt{3} & 0 \\ 0 & 0 & 0 & -1 \\ 0 & 0 & 0 & 0 \end{pmatrix} \quad e(3,0) = \frac{1}{\sqrt{20}}\begin{pmatrix} 1 & 0 & 0 & 0 \\ 0 & -3 & 0 & 0 \\ 0 & 0 & 3 & 0 \\ 0 & 0 & 0 & -1 \end{pmatrix}$$

$$e(2,2) = \frac{1}{\sqrt{2}} \begin{pmatrix} 0 & 0 & 1 & 0 \\ 0 & 0 & 0 & 1 \\ 0 & 0 & 0 & 0 \\ 0 & 0 & 0 & 0 \end{pmatrix} \quad e(2,1) = \frac{1}{\sqrt{2}} \begin{pmatrix} 0 & -1 & 0 & 0 \\ 0 & 0 & 0 & 0 \\ 0 & 0 & 0 & 1 \\ 0 & 0 & 0 & 0 \end{pmatrix} \quad e(2,0) = \frac{1}{2} \begin{pmatrix} 1 & 0 & 0 & 0 \\ 0 & -1 & 0 & 0 \\ 0 & 0 & -1 & 0 \\ 0 & 0 & 0 & 1 \end{pmatrix}$$

$$e(1,1) = \frac{-1}{\sqrt{10}} \begin{pmatrix} 0 & \sqrt{3} & 0 & 0 \\ 0 & 0 & 2 & 0 \\ 0 & 0 & 0 & \sqrt{3} \\ 0 & 0 & 0 & 0 \end{pmatrix} \quad e(1,0) = \frac{1}{\sqrt{20}} \begin{pmatrix} 3 & 0 & 0 & 0 \\ 0 & 1 & 0 & 0 \\ 0 & 0 & -1 & 0 \\ 0 & 0 & 0 & -3 \end{pmatrix}$$

Remark 3.3.11. Let us also make the observation that all non-zero Clebsch-Gordan coefficients (3.39) with $j_1 = j_2$ can be read off from the entries of the matrices $e^j(l,m)$, namely by (3.32) and (3.44) there is the expansion

$$e^j(l,m) = \sum_k (-1)^{m+k-1} C_{j-k+1,m-j+k-1,m}^{j,j,l} \mathcal{E}_{k,m+k} \tag{3.82}$$

where the summation is in the range

$$\begin{cases} k = 1,2,\ldots,n+1-m, & \text{when } m \geq 0, \\ k = |m|+1, |m|+2, \ldots, n+1, & \text{when } m < 0. \end{cases}$$

Therefore, the m-subdiagonal matrix (3.82), presented as in (3.60), is

$$e^j(l,m) = (e_1, e_2, \ldots, e_{n+1-|m|})_m \tag{3.83}$$

where

$$e_k = \begin{cases} (-1)^{m+k-1} C_{j-k+1,m-j+k-1,m}^{j,j,l}, & \text{when } m \geq 0, \\ (-1)^{k-1} C_{j-|m|-k+1,-j+k-1, m}^{j,j,l}, & \text{when } m < 0. \end{cases} \tag{3.84}$$

Let us illustrate this for $j = 1$ (see above), where there are 17 coefficients $C_{m_1,m_2,m}^{1,1,l}$ for the appropriate range of indices, namely

$$C_{m_1,m_2,m}^{1,1,l}, \quad |m_i| \leq 1, m = m_1 + m_2, |m| \leq l \leq 2.$$

Thus, for example,

$$e(1,1) = (-1/\sqrt{2}, -1\sqrt{2})_1 = (-C_{1,0,1}^{1,1,1}, C_{0,1,1}^{1,1,1})_1,$$

$$e(1,0) = (1/\sqrt{2}, 0, -1/\sqrt{2})_0 = (C_{1,-1,0}^{1,1,1}, -C_{0,0,0}^{1,1,1}, C_{-1,1,0}^{1,1,1})_0,$$

$$e(0,0) = (1/\sqrt{3}, 1/\sqrt{3}, 1/\sqrt{3})_0 = (C_{1,-1,0}^{1,1,0}, -C_{0,0,0}^{1,1,0}, C_{-1,1,0}^{1,1,0})_0$$

and among the above 17 coefficients only $C_{0,0,0}^{1,1,1}$ vanishes.

3.3.3 Decomposition of the Operator Product

Linearly, the algebra $M_\mathbb{C}(n+1)$ is spanned by the matrices $\mathbf{e}(l,m)$, so it is natural to inquire about their multiplication laws. But these are normalized matrices, namely the scaled version of the matrices $E(l,m)$, cf. (3.72).

The Parity Property

Let us first state a *parity property*, to be established later, for the commutator $[P,Q] = PQ-QP$ and anti-commutator $[[P,Q]] = PQ+QP$ of these unnormalized basis vectors $E(l,m)$, as follows:

Proposition 3.3.12 (The Parity Property for operators). *The following multiplication rules hold for the matrices* $E(l,m)$:

$$(i): [E(l_1,m_1), E(l_2,m_2)] = \sum_{l \equiv l_1+l_2+1} K_{m_1,m_2}^{l_1,l_2,l} E(l,m_1+m_2), \qquad (3.85)$$

$$(ii): [[E(l_1,m_1), E(l_2,m_2)]] = \sum_{l \equiv l_1+l_2} K_{m_1,m_2}^{l_1,l_2,l} E(l,m_1+m_2), \qquad (3.86)$$

with sums restricted by $\delta(l_1,l_2,l) = 1$, *and* $l \equiv k$ *means congruence modulo* 2.

Remark 3.3.13. The importance of commutators in quantum mechanics stemmed from Heisenberg-Born-Jordan's matrix mechanics. Given a preferred Hermitian matrix H, called the Hamiltonian matrix, it generates a flow on Hilbert space defining the dynamics of a time-dependent matrix M by *Heisenberg's equation*:

$$\frac{dM}{dt} = \frac{1}{i\hbar}[H,M] + \frac{\partial M}{\partial t}, \qquad (3.87)$$

where $\hbar$ is Planck's constant, and we emphasize that the Hamiltonian matrix H is always a time-independent (constant) Hermitian matrix.

When M does not depend explicitly on time, Eq. (3.87) is the time derivative of the map (Adjoint action, or action by conjugation, cf. (ii) in (3.54))

$$M \mapsto \exp(-itH/\hbar) M \exp(itH/\hbar) \qquad (3.88)$$

and the one-parameter subgroup of $U(n+1)$ given by

$$\Phi_H = \{\exp(-itH/\hbar), t \in \mathbb{R}\}$$

determines the flow of H on $\mathcal{H}_j$ which defines the dynamics of the spin-j system. By allowing the entries of M to have an extra, explicit dependence on time, one obtains (3.87) as the time derivative of the r.h.s. of (3.88).

We also recall that, in accordance with the first part of (3.87), commutators act as derivations on the algebra of operators, that is, $\forall H, M, N \in M_{\mathbb{C}}(n+1)$,

$$[H, MN] = M[H, N] + [H, M]N .\tag{3.89}$$

Finally, we remark that Eq. (3.87) generalizes to contexts other than spin-j systems, as in affine mechanical systems, to define quantum dynamics of bounded operators on other (finite or infinite dimensional) Hilbert spaces.

Remark 3.3.14. For a fixed n, the multi-indexed coefficients K in the expansions (3.85) and (3.86) are seen to be rational numbers. In fact, they are polynomials in n whose degree increases stepwise by 2 as l decreases by 2.

Example 3.3.15. For $n \geq 5$,

$$E(3, 2)E(2, 1) = k_3 E(3, 3) + k_4 E(4, 3) + k_5 E(5, 3),$$

$$E(2, 1)E(3, 2) = k_3 E(3, 3) - k_4 E(4, 3) + k_5 E(5, 3)$$

where $k_3 = \frac{2}{3}n^2 + \frac{4}{3}n - 22$, $k_4 = 3/2$, and $k_5 = 4/15$.

We shall give two proofs of Proposition 3.3.12; one at the end of this chapter, a corollary of the full product rule to be developed below, and an independent one in Appendix 10.2.

The Full Product Rule

Indeed, it is possible to obtain in a very straightforward way the full multiplication rule for the standard coupled basis vectors given by Theorem 3.3.6, conveniently denoted in different ways such as

$$\mathbf{e}(l, m) = \mathbf{e}^j (l, m) = |(jj)lm\rangle \ , \quad \langle (jj)l'm'|(jj)lm\rangle = \delta_{l,l'}\delta_{m,m'},$$

from their Clebsch-Gordan expansions in terms of the uncoupled basis vectors $|j_1 m_1 j_2 m_2\rangle$. Thus, from Definition 3.2.1 (model (ii)) and Eqs. (3.40) and (3.44), together with the multiplication rule for one-element matrices

$$\mathcal{E}_{i,j}\mathcal{E}_{k,l} = \delta_{j,k}\mathcal{E}_{i,l},\tag{3.90}$$

we are straightforwardly led to the following result:

Theorem 3.3.16. *The operator product of the standard coupled basis vectors decomposes in the standard coupled basis according to the following formula:*

$$\mathbf{e}^j(l_1, m_1)\mathbf{e}^j(l_2, m_2) = \sum_{l=0}^{2j} \mathcal{M}[j]_{m_1,m_2,m}^{l_1,\, l_2,\, l}\mathbf{e}^j(l, m) \tag{3.91}$$

where $m = m_1 + m_2$ and the product coefficients can be expressed as

$$\mathcal{M}[j]_{m_1,m_2,m}^{l_1,\, l_2,\, l} = \sum_{\mu_1=-j}^{j} \sum_{\mu_2=-j}^{j} \sum_{\mu_3=-j}^{j} (-1)^{j+\mu_2} C_{\mu_1,\mu_2,m_1}^{j,\, j,\, l_1}\, C_{-\mu_2,\mu_3,m_2}^{j,\, j,\, l_2}\, C_{\mu_1,\mu_3,m}^{j,\, j,\, l}. \tag{3.92}$$

Remark 3.3.17. At first sight, there should be a summation in m from $-l$ to l in Eq. (3.91). However, it follows straightforwardly from Eq. (3.92) and the non-vanishing conditions for the Clebsch-Gordan coefficients (cf. (3.41)), that the *product coefficients* $\mathcal{M}[j]$ vanish unless $m = m_1 + m_2$, implying no summation in m in Eq. (3.91), in agreement with Eqs. (3.85) and (3.86).

In fact, as we shall see more clearly below (cf. Eqs. (3.97) and (3.111)), the product coefficients satisfy non-vanishing conditions similar to those of the Clebsch-Gordan coefficients, (cf. (3.41)), namely

$$\mathcal{M}[j]_{m_1,m_2,m}^{l_1,\, l_2,\, l} \neq 0 \implies \begin{cases} m = m_1 + m_2 \\ \delta(l_1, l_2, l) = 1 \end{cases}, \tag{3.93}$$

and clearly $\delta(j, j, l_i) = \delta(j, j, l) = 1$ also holds.

Wigner Symbols and the Product Rule

It is convenient to introduce the *Wigner 3jm symbols* (cf. [74]), which are closely related to the Clebsch-Gordan coefficients.

Definition 3.3.18. The *Wigner 3jm symbol* is defined to be the rightmost symbol of the identity

$$C_{m_1,m_2,-m_3}^{j_1,j_2,j_3} = (-1)^{j_1-j_2-m_3}\sqrt{2j_3+1}\begin{pmatrix} j_1 & j_2 & j_3 \\ m_1 & m_2 & m_3 \end{pmatrix}. \tag{3.94}$$

As a substitute for the Clebsch-Gordan coefficients, the relevance of the Wigner 3jm symbols is largely due to their better symmetry properties:

$$\begin{pmatrix} j_1 & j_2 & j_3 \\ m_1 & m_2 & m_3 \end{pmatrix} = \begin{pmatrix} j_2 & j_3 & j_1 \\ m_2 & m_3 & m_1 \end{pmatrix} = \begin{pmatrix} j_3 & j_1 & j_2 \\ m_3 & m_1 & m_2 \end{pmatrix} \tag{3.95}$$

$$= (-1)^{j_1+j_2+j_3}\begin{pmatrix} j_2 & j_1 & j_3 \\ m_2 & m_1 & m_3 \end{pmatrix} = (-1)^{j_1+j_2+j_3}\begin{pmatrix} j_1 & j_2 & j_3 \\ -m_1 & -m_2 & -m_3 \end{pmatrix}$$

which are obtained directly from (3.94) and the symmetry properties (3.50)–(3.52) for the Clebsch-Gordan coefficients. Furthermore, the non-vanishing conditions for the Wigner $3jm$ symbols analogous to (3.41) are more symmetric:

$$
\begin{pmatrix} j_1 & j_2 & j_3 \\ m_1 & m_2 & m_3 \end{pmatrix} \neq 0 \;\Rightarrow\; \begin{cases} m_1 + m_2 + m_3 = 0 \\ \delta(j_1, j_2, j_3) = 1 \end{cases}
\tag{3.96}
$$

Thus, let us also introduce another symbol, in analogy with (3.94).

Definition 3.3.19. The *Wigner product symbol* is defined to be the rightmost symbol of the identity

$$
\mathcal{M}[j]_{m_1,m_2,-m_3}^{l_1,l_2,l_3} = (-1)^{2j-m_3} \begin{bmatrix} l_1 & l_2 & l_3 \\ m_1 & m_2 & m_3 \end{bmatrix}[j].
\tag{3.97}
$$

From Eqs. (3.92), (3.94) and (3.97), the *Wigner product symbol* is also defined by its relation to the Wigner $3jm$ symbols according to the identity

$$
\begin{bmatrix} l_1 & l_2 & l_3 \\ m_1 & m_2 & m_3 \end{bmatrix}[j]
\tag{3.98}
$$

$$
= \sqrt{(2l_1+1)(2l_2+1)(2l_3+1)} \sum_{\mu_1,\mu_2,\mu_3=-j}^{j} (-1)^{3j-\mu_1-\mu_2-\mu_3}
$$

$$
\cdot \begin{pmatrix} j & l_1 & j \\ \mu_1 & m_1 & -\mu_2 \end{pmatrix} \begin{pmatrix} j & l_2 & j \\ \mu_2 & m_2 & -\mu_3 \end{pmatrix} \begin{pmatrix} j & l_3 & j \\ \mu_3 & m_3 & -\mu_1 \end{pmatrix}.
$$

With the above definition, Eq. (3.91) can be restated as:

$$
\mathbf{e}^j(l_1,m_1)\mathbf{e}^j(l_2,m_2) = \sum_{l=0}^{2j} (-1)^{2j+m} \begin{bmatrix} l_1 & l_2 & l \\ m_1 & m_2 & -m \end{bmatrix}[j]\, \mathbf{e}^j(l,m).
\tag{3.99}
$$

In order to re-interpret the Wigner product symbol defined by Eq. (3.98) via a sum of triple products of Wigner $3jm$ symbols, it is convenient to introduce another kind of Wigner symbol already studied in the literature. These are the Wigner $6j$ symbols, which are related to the re-coupling coefficients that appear when taking triple tensor products of irreducible $SU(2)$-representations.

For simplicity, let us refer to three representations $\varphi_{j_1}, \varphi_{j_2}, \varphi_{j_3}$, as j_1, j_2, j_3, respectively. Thus, if $|j_1 m_1 j_2 m_2 j_3 m_3\rangle$ is an uncoupled basis vector of the triple tensor product of j_1, j_2, j_3, a coupled basis vector for the tensor product can be obtained by first coupling j_1 and j_2 and then coupling with j_3, or first coupling j_2 and j_3 and then coupling with j_1, or still, first coupling j_3 and j_1 and then with j_2. Symbolically we describe the three coupling schemes by

$$(i) \quad j_1 + j_2 = j_{12}, \quad j_{12} + j_3 = j,$$

$$(ii) \quad j_2 + j_3 = j_{23}, \quad j_{23} + j_1 = j,$$

$$(iii) \quad j_3 + j_1 = j_{31}, \quad j_{31} + j_2 = j$$

and the coupled basis vectors arising from the three coupling schemes (i), (ii) and (iii), respectively, are given by

$$|(j_{12}j_3)jm\rangle = \sum_{m_1=-j_1}^{j_1} \sum_{m_2=-j_2}^{j_2} \sum_{m_3=-j_3}^{j_3} C_{m_1,m_2,m_{12}}^{j_1,j_2,j_{12}} C_{m_{12},m_3,m}^{j_{12},j_3,j} |j_1 m_1 j_2 m_2 j_3 m_3\rangle,$$

$$|(j_{23}j_1)jm\rangle = \sum_{m_1=-j_1}^{j_1} \sum_{m_2=-j_2}^{j_2} \sum_{m_3=-j_3}^{j_3} C_{m_2,m_3,m_{23}}^{j_2,j_3,j_{23}} C_{m_{23},m_1,m}^{j_{23},j_1,j} |j_1 m_1 j_2 m_2 j_3 m_3\rangle,$$

$$|(j_{31}j_2)jm\rangle = \sum_{m_1=-j_1}^{j_1} \sum_{m_2=-j_2}^{j_2} \sum_{m_3=-j_3}^{j_3} C_{m_3,m_1,m_{31}}^{j_3,j_1,j_{31}} C_{m_{31},m_2,m}^{j_{31},j_2,j} |j_1 m_1 j_2 m_2 j_3 m_3\rangle,$$

$$(3.100)$$

where in case (i), for example, the left side is a coupled basis vector arising from the tensor product of Hilbert spaces with standard basis $\{|(j_1 j_2) j_{12} m_{12})\}$ and $\{|j_3 m_3\rangle\}$. The range of j_{12} is determined by the condition $\delta(j_1, j_2, j_{12}) = 1$.

Definition 3.3.20. The inner products between these coupled basis vectors are the *re-coupling coefficients* and they define the *Wigner 6j symbol* as the rightmost symbol of the identity

$$\langle (j_{12}j_3)jm|(j_{23}j_1)jm \rangle \tag{3.101}$$

$$= (-1)^{j_1+j_2+j_3+j} \sqrt{(2j_{12}+1)(2j_{23}+1)} \begin{Bmatrix} j_1 & j_2 & j_{12} \\ j_3 & j & j_{23} \end{Bmatrix}.$$

Therefore, using (3.100), the *Wigner 6j symbols* can be written as

$$\begin{Bmatrix} j_1 & j_2 & j_{12} \\ j_3 & j & j_{23} \end{Bmatrix} = \frac{(-1)^{j_1+j_2+j_3+j}}{\sqrt{(2j_{12}+1)(2j_{23}+1)}} \tag{3.102}$$

$$\cdot \sum_{m_1=-j_1}^{j_1} \sum_{m_2=-j_2}^{j_2} \sum_{m_3=-j_3}^{j_3} C_{m_1,m_2,m_{12}}^{j_1,j_2,j_{12}} C_{m_{12},m_3,m}^{j_{12},j_3,j} C_{m_2,m_3,m_{23}}^{j_2,j_3,j_{23}} C_{m_{23},m_1,m}^{j_{23},j_1,j}$$

with similar equations for the Wigner $6j$ symbols obtained by the re-coupling coefficients $\langle (j_{23}j_1)jm|(j_{31}j_2)jm \rangle$ and $\langle (j_{31}j_2)jm|(j_{12}j_3)jm \rangle$.

Replacement of the Clebsch-Gordan coefficients in (3.102) by the Wigner 3jm symbols using Eq. (3.94) yields the following symmetric expression for the Wigner $6j$ symbols (cf. [74]):

$$\begin{Bmatrix} a & b & c \\ d & e & f \end{Bmatrix} = \sum (-1)^{d+e+f+\delta+\epsilon+\phi}. \tag{3.103}$$

$$\cdot \begin{pmatrix} a & b & c \\ \alpha & \beta & \gamma \end{pmatrix} \begin{pmatrix} a & e & f \\ \alpha & \epsilon & -\phi \end{pmatrix} \begin{pmatrix} d & b & f \\ -\delta & \beta & \phi \end{pmatrix} \begin{pmatrix} d & e & c \\ \delta & -\epsilon & \gamma \end{pmatrix}$$

where the sum is taken over all possible values of $\alpha, \beta, \gamma, \delta, \epsilon, \phi$, remembering that only three of these are independent. For example, since the sum of the numbers in the second row of a $3jm$ symbol is zero, we have the relations

$$\alpha = -\epsilon + \phi, \beta = \delta - \phi, \gamma = \epsilon - \delta.$$

We should note that the six numbers in a Wigner $6j$ symbol are on an equal footing, representing total spins and not projection quantum number m_i.

Now, we shall use an identity (cf. [74]) which can be obtained from the orthonormality relation for Wigner $3jm$ symbols. The latter is derived from the one for Clebsch-Gordan coefficients (3.43) and combined with the symmetry properties (3.95) of the Wigner $3jm$ symbols one obtains the identity

$$\begin{pmatrix} a & b & c \\ -\alpha & -\beta & -\gamma \end{pmatrix} \begin{Bmatrix} a & b & c \\ d & e & f \end{Bmatrix}$$

$$= \sum_{\delta,\epsilon,\phi} (-1)^{d-\delta+e-\epsilon+f-\phi} \begin{pmatrix} e & a & f \\ \epsilon & \alpha & -\phi \end{pmatrix} \begin{pmatrix} f & b & d \\ \phi & \beta & -\delta \end{pmatrix} \begin{pmatrix} d & c & e \\ \delta & \gamma & -\epsilon \end{pmatrix}.$$

Together with Eq. (3.98), this yields the following result:

Proposition 3.3.21. *The Wigner product symbol is proportional to the product of a Wigner $3jm$ symbol and a Wigner $6j$ symbol, precisely as*

$$\begin{bmatrix} l_1 & l_2 & l_3 \\ m_1 & m_2 & m_3 \end{bmatrix} [j] \tag{3.104}$$

$$= \sqrt{(2l_1 + 1)(2l_2 + 1)(2l_3 + 1)} \begin{pmatrix} l_1 & l_2 & l_3 \\ -m_1 & -m_2 & -m_3 \end{pmatrix} \begin{Bmatrix} l_1 & l_2 & l_3 \\ j & j & j \end{Bmatrix}.$$

Then, from Eqs. (3.99) and (3.104) we have immediately:

Corollary 3.3.22. *The operator product of the standard coupled basis vectors $\mathbf{e}^j (l, m)$, stated in Theorem 3.3.16, is given by*

$$\mathbf{e}^j (l_1, m_1) \mathbf{e}^j (l_2, m_2) = \sum_{l=0}^{2j} (-1)^{2j+m} \sqrt{(2l_1 + 1)(2l_2 + 1)(2l + 1)}$$

$$\cdot \begin{pmatrix} l_1 & l_2 & l \\ -m_1 & -m_2 & m \end{pmatrix} \begin{Bmatrix} l_1 & l_2 & l \\ j & j & j \end{Bmatrix} \mathbf{e}^j (l, m).$$

Remark 3.3.23.   (i) By linearity, given operators, $F = \sum_{l=0}^{n} \sum_{m=-l}^{l} F_{lm} \mathbf{e}^j (l, m)$ and $G = \sum_{l=0}^{n} \sum_{m=-l}^{l} G_{lm} \mathbf{e}^j (l, m)$, with $F_{lm}, G_{lm} \in \mathbb{C}$, their operator product decomposes as $FG = \sum_{l=0}^{n} \sum_{m=-l}^{l} (FG)_{lm} \mathbf{e}^j (l, m)$, with $(FG)_{lm} \in \mathbb{C}$ given by

$$(FG)_{lm} = (-1)^{n+m} \sum_{l_1, l_2 = 0}^{n} \sum_{m_1 = -l_1}^{l_1} \begin{bmatrix} l_1 & l_2 & l \\ m_1 & m_2 & -m \end{bmatrix} [j] F_{l_1 m_1} G_{l_2 m_2}$$

where $m_2 = m - m_1$ and the sum in l_1, l_2 is restricted by $\delta(l_1, l_2, l) = 1$.

(ii) One should compare the above equation with the equation for the product of operators F and G decomposed in the orthonormal basis of one-element matrices $\mathcal{E}_{i,j}$ so that, for $F = \sum_{i,j=1}^{n+1} F_{ij} \mathcal{E}_{i,j}$, $G = \sum_{i,j=1}^{n+1} G_{ij} \mathcal{E}_{i,j}$, $F_{ij}, G_{ij} \in \mathbb{C}$, from (3.90) one has the usual much simpler expression for the matrix product

$$(FG)_{ij} = \sum_{k=1}^{n+1} F_{ik} G_{kj} .$$

Of course, the problem with this familiar and simple product decomposition is that it is not $SO(3)$-invariant. However, our main justification for going through the $SO(3)$-invariant decomposition of the operator product in the coupled basis $\mathbf{e}^j (l, m)$ will become clear later on, in Chap. 7.

Explicit Formulae

Now, in view of the above remark, it is interesting to have some explicit expressions for the Wigner product symbol. However, from Proposition 3.3.21, this amounts to having explicit formulae for the Wigner $3jm$ symbol and the particular Wigner $6j$ symbol appearing in Eq. (3.104).

The first ones are obtained from (3.94) and the expressions for the Clebsch-Gordan coefficients. For completeness, we list here the one obtained from (3.47):

$$\begin{pmatrix} l_1 & l_2 & l_3 \\ m_1 & m_2 & m_3 \end{pmatrix} = \Delta(l_1, l_2, l_3) \, S_{m_1, m_2, m_3}^{l_1, l_2, l_3} \, N_{m_1, m_2, m_3}^{l_1, l_2, l_3} , \qquad (3.105)$$

where $\Delta(l_1, l_2, l_3)$ and $S_{m_1, m_2, m_3}^{l_1, l_2, l_3}$ are given respectively by (3.48) and (3.49) via substitution of j_i for l_i and

$$N_{m_1, m_2, m_3}^{l_1, l_2, l_3} = \qquad\qquad\qquad\qquad\qquad\qquad\qquad\qquad\qquad\qquad (3.106)$$

$$\sum_z \frac{(-1)^{l_1 - l_2 - m_3 + z}}{z!(l_1 + l_2 - l_3 - z)!(l_1 - m_1 - z)!(l_2 + m_2 - z)!(l_3 - l_2 + m_1 + z)!(l_3 - l_1 - m_2 + z)!}$$

with the usual summation convention (cf. Remark 3.2.8).

We recall that Eq. (3.105) holds only under the non-vanishing conditions (3.96) for the Wigner $3jm$ symbols. Also, it is immediate from Eqs. (3.48) and (3.49) that $\Delta(l_1, l_2, l_3)$ and $S_{m_1,m_2,m_3}^{l_1,\ l_2,\ l_3}$ are invariant under any permutation of the columns in the Wigner $3jm$ symbol and any change in sign of the magnetic numbers m_i. Therefore, it is the latter function $N_{m_1,m_2,m_3}^{l_1,\ l_2,\ l_3}$ that carries the symmetry properties of the Wigner $3jm$ symbol, namely,

$$N_{m_1,m_2,m_3}^{l_1,\ l_2,\ l_3} = N_{m_3,m_1,m_2}^{l_3,\ l_1,\ l_2} = (-1)^{l_1+l_2+l_3} N_{m_2,m_1,m_3}^{l_2,\ l_1,\ l_3} \tag{3.107}$$

$$= (-1)^{l_1+l_2+l_3} N_{-m_1,-m_2,-m_3}^{l_1,\ l_2,\ l_3} \ .$$

Explicit expressions for the Wigner $6j$ symbol in Eq. (3.104) can be obtained as a particular case of the known explicit expressions for general Wigner $6j$ symbols which have been obtained from equations like (3.102) and (3.103) and are listed (cf. e.g. [13, 74]). The expression obtained by Racah in 1942 yields:

$$\left\{ \begin{array}{ccc} l_1 & l_2 & l_3 \\ j & j & j \end{array} \right\} = l_1! l_2! l_3! \Delta(l_1, l_2, l_3) \sqrt{\frac{(n-l_1)!(n-l_2)!(n-l_3)!}{(n+l_1+1)!(n+l_2+1)!(n+l_3+1)!}}$$

$$\cdot \sum_k \frac{(-1)^{n+k}(n+1+k)!}{(n+k-l_1-l_2-l_3)! R(l_1, l_2, l_3; k)} \ , \tag{3.108}$$

where $n = 2j$, and

$$R(l_1, l_2, l_3; k) \tag{3.109}$$

$$= (k-l_1)!(k-l_2)!(k-l_3)!(l_1+l_2-k)!(l_2+l_3-k)!(l_3+l_1-k)! \ \ .$$

We note that the function $\Delta(l_1, l_2, l_3)$ given by (3.48) appears in both expressions (3.105) and (3.108). Therefore, by Proposition 3.3.21 and Eq. (3.104) this function appears squared in the expression for the Wigner product symbol.

Symmetry Properties of the Product Rule

Now, the following is immediate from (3.108)–(3.109):

Proposition 3.3.24. $\left\{ \begin{array}{ccc} l_1 & l_2 & l_3 \\ j & j & j \end{array} \right\}$ *is invariant by any permutation of* $(l_1 l_2 l_3)$.

This property can also be obtained directly by algebraic manipulations starting from Eq. (3.103), as shown in Appendix 10.3.

Remark 3.3.25. In fact, Proposition 3.3.24 is a particular case of the more general statement: every Wigner $6j$ symbol is invariant under any permutation of its columns and under any exchange of the upper and lower numbers in any given column. A proof of this statement, which follows from the symmetries of the $3jm$ symbols and the associativity of the triple tensor product, is found in [13].

From Proposition 3.3.24 and Eq. (3.104) in Proposition 3.3.21, we have:

Corollary 3.3.26. *The Wigner product symbols (3.98) have the same symmetry and non-vanishing properties as the Wigner $3jm$ symbols, namely (cf. (3.95)),*

$$
\begin{bmatrix} l_1 & l_2 & l_3 \\ m_1 & m_2 & m_3 \end{bmatrix}[j] = \begin{bmatrix} l_2 & l_3 & l_1 \\ m_2 & m_3 & m_1 \end{bmatrix}[j] = \begin{bmatrix} l_3 & l_1 & l_2 \\ m_3 & m_1 & m_2 \end{bmatrix}[j] \tag{3.110}
$$

$$
= (-1)^{l_1+l_2+l_3} \begin{bmatrix} l_2 & l_1 & l_3 \\ m_2 & m_1 & m_3 \end{bmatrix}[j] = (-1)^{l_1+l_2+l_3} \begin{bmatrix} l_1 & l_2 & l_3 \\ -m_1 & -m_2 & -m_3 \end{bmatrix}[j],
$$

$$
\begin{bmatrix} l_1 & l_2 & l_3 \\ m_1 & m_2 & m_3 \end{bmatrix}[j] \neq 0 \quad \Rightarrow \quad \begin{cases} m_1 + m_2 + m_3 = 0 \\ \delta(l_1, l_2, l_3) = 1 \end{cases} . \tag{3.111}
$$

Proof of the parity property: Together with Eq. (3.99), the identities (3.110) and (3.111) imply the parity property for operators, as stated in Proposition 3.3.12.

Remark 3.3.27. By comparison with the direct proof of Proposition 3.3.12 presented in Appendix 10.2, a look at the above proof is enough to indicate the great amount of combinatorics that is encoded by the Wigner $3jm$ and $6j$ symbols. We refer to [13, 74] for overviews of their further properties.

Chapter 4
The Poisson Algebra of the Classical Spin System

This chapter presents the basic mathematical framework for classical mechanics of a spin system. Practically all the material in the introductory section below can be found in basic textbooks on classical mechanics and we refer to some of these, e.g. [1, 5, 34, 37, 48], for the reader not yet too familiar with the subject, or for further details, examples, etc. (Ref. [34] is more familiar to physicists, while the others are more mathematical and closer in style to our brief introduction below). Our emphasis here is to provide a self-contained presentation of the $SO(3)$-invariant decomposition of the pointwise product and the Poisson bracket of polynomials, which are not easily found elsewhere (specially the latter).

4.1 Basic Definitions of the Classical Spin System

In this section we collect some basic facts and definitions concerning the Poisson algebra of functions on the 2-sphere $G/H = S^2$, whose homogeneous space structure was exploited in (2.38), namely $G = SO(3)$ or $SU(2)$ acts by rotations, and in the latter case G acts via a (fixed) covering homomorphism (cf. (2.23) and (2.32)):

$$\psi : SU(2) \rightarrow SO(3).$$

We recall that the homogeneous 2-sphere carries a G-invariant symplectic (or area) form ω, cf. also (2.18), which is (locally) expressed in terms of spherical polar coordinates (2.39), measured with respect to the "north pole" $\mathbf{n}_0$, as in (2.40), that is,

$$\omega = \sin \varphi d\varphi \wedge d\theta . \tag{4.1}$$

In other words, the 2-sphere is a particular case of the following:

© Springer International Publishing Switzerland 2014
P. de M. Rios, E. Straume, *Symbol Correspondences for Spin Systems*,
DOI 10.1007/978-3-319-08198-4_4

Definition 4.1.1. A *symplectic manifold* (M, ω) is a smooth manifold M endowed with a closed nondegenerate two-form ω, which is called a *symplectic form*.

Remark 4.1.2. It follows from the definition that the dimension of a real symplectic manifold M is an even number $2d$. If a mechanical system is modeled on M (see more below), then $d \in \mathbb{N}$ is called the number of degrees of freedom of the system.

Remark 4.1.3. We often write $\omega = dS$ to indicate that ω is the surface element, but we emphasize that this is a shorthand notation: ω is not an exact form. The local expression (4.1) for the $SO(3)$-invariant symplectic form ω on the sphere is canonical up to a choice of orientation, or sign $\pm$. For the standard choice of orientation on $\mathbb{R}^3 = SO(3) = SU(2)$, $dx \wedge dy \wedge dz > 0$, the induced orientation for ω is via the identification

$$\omega = (xdy \wedge dz + ydz \wedge dx + zdx \wedge dy)_{x^2+y^2+z^2=1} \tag{4.2}$$

and is the one with the choice of $+$ sign given by Eq. (4.1). Formula (4.2) above also provides a direct way to verify that ω is symplectic and G-invariant.

For a complex-valued continuous function on S^2, its normalized integral over S^2 equals its integral over G, namely

$$\int_G F(g\mathbf{n}_0)dg = \frac{1}{4\pi} \int_{S^2} F(\mathbf{n})dS = \frac{1}{4\pi} \int_0^{2\pi} \int_0^{\pi} F(\theta, \varphi) \sin \varphi d\varphi d\theta \tag{4.3}$$

where dg denotes the normalized Haar integral, and $\mathbf{n}_0 \in S^2$ is the "north pole" fixed by $H = SO(2)$, cf. (2.37). For functions F_i on S^2 the L^2-inner product is

$$\langle F_1, F_2 \rangle = \frac{1}{4\pi} \int_{S^2} \overline{F_1(\mathbf{n})} F_2(\mathbf{n})dS, \tag{4.4}$$

in particular, the constant 1 has norm 1. On the other hand, the metric on S^2 and the gradient of a function f are given in local spherical coordinates by

$$ds^2 = d\varphi^2 + \sin^2 \varphi d\theta^2, \quad \nabla f = \frac{\partial f}{\partial \varphi} \frac{\partial}{\partial \varphi} + \frac{1}{\sin^2 \varphi} \frac{\partial f}{\partial \theta} \frac{\partial}{\partial \theta}. \tag{4.5}$$

Besides the ordinary pointwise multiplication of functions on the sphere, which is commutative, another classical product of smooth functions on the sphere is defined using the symplectic form and this turns out to be anti-commutative, or skew-symmetric.

In what follows, if α is an n-form and v is a vector field, let $v \lrcorner \alpha$ denote the $(n-1)$-form obtained via interior product of v and α. Then, we have the following:

Definition 4.1.4. For any smooth function f on S^2, its *Hamiltonian vector field* X_f is defined by

$$X_f \lrcorner \, \omega + df = 0 \, . \tag{4.6}$$

In local spherical coordinates, the Hamiltonian vector field has the expression

$$X_f = \frac{1}{\sin\varphi} \left(\frac{\partial f}{\partial\varphi} \frac{\partial}{\partial\theta} - \frac{\partial f}{\partial\theta} \frac{\partial}{\partial\varphi} \right) \, . \tag{4.7}$$

Definition 4.1.5. The *Poisson bracket* of two smooth functions is, by definition,

$$\{f_1, f_2\} = X_{f_1}(f_2) = -X_{f_2}(f_1) = \omega(X_{f_1}, X_{f_2}) \, . \tag{4.8}$$

In local spherical coordinates, it is written as

$$\{f_1, f_2\} = \frac{1}{\sin\varphi} \left(\frac{\partial f_1}{\partial\varphi} \frac{\partial f_2}{\partial\theta} - \frac{\partial f_1}{\partial\theta} \frac{\partial f_2}{\partial\varphi} \right) \, . \tag{4.9}$$

In particular, it follows that

$$\{x, y\} = z, \{y, z\} = x, \{z, x\} = y \tag{4.10}$$

and, furthermore, it follows immediately from (4.8):

Proposition 4.1.6. *The Poisson bracket is a derivation with respect to the ordinary pointwise product of functions:*

$$\{f_1, f_2 f_3\} = \{f_1, f_2\} f_3 + f_2 \{f_1, f_3\} \, . \tag{4.11}$$

Finally, while the ordinary commutative product is associative, for the skew-symmetric Poisson bracket we have:

Proposition 4.1.7. *The Poisson bracket satisfies the Jacobi identity:*

$$\{\{f_1, f_2\}, f_3\} + \{\{f_2, f_3\}, f_1\} + \{\{f_3, f_1\}, f_2\} = 0 \tag{4.12}$$

and thus defines a Lie algebra on the space of smooth functions on the sphere.

Proof. First, note that for any Hamiltonian vector field X_f defined by (4.6), using Cartan's "magic" formula

$$\mathcal{L}_{X_f}\omega = X_f \lrcorner \, d\omega + d(X_f \lrcorner \, \omega) \, ,$$

it follows from closedness of ω, $d\omega = 0$, and from (4.6) that

$$\mathcal{L}_{X_f}\omega = 0 . \tag{4.13}$$

But then it follows from (4.8) that

$$\mathcal{L}_{X_f}(\{g, h\}) = X_f(\{g, h\}) = \{X_f(g), h\} + \{g, X_f(h)\} .$$

And from (4.8) this equation above is equivalent to

$$\{f, \{g, h\}\} = \{\{f, g\}, h\} + \{g, \{f, h\}\} ,$$

which is equivalent to (4.12).   □

Corollary 4.1.8. *The Jacobi identity (4.12) is equivalent to*

$$[X_f, X_g] = X_{\{f,g\}} . \tag{4.14}$$

Proof. By definition,

$$
\begin{aligned}
[X_f, X_g](h) &= X_f(X_g(h)) - X_g(X_f(h)) \\
&= X_f(\{g, h\}) - X_g(\{f, h\}) \\
&= \{f, \{g, h\}\} - \{g, \{f, h\}\} \\
&= \{\{f, g\}, h\} \\
&= X_{\{f,g\}}(h) ,
\end{aligned}
$$

where we have repeatedly used (4.8) and once used (4.12). And similarly, from (4.14) we get (4.12).   □

Because of (4.12)–(4.14), the Poisson algebra is also called the Poisson-Lie algebra of smooth functions on a symplectic manifold.

Definition 4.1.9. The *Poisson algebra* of S^2 is the space of smooth complex functions on S^2 with its commutative pointwise product $\cdot$ and anti-commutative Poisson bracket $\{\ ,\ \}$ defined by the $SO(3)$-invariant symplectic form ω given by (4.1) and (4.2), via (4.6)–(4.9), satisfying (4.11) and (4.12), which shall be denoted by $\{\mathcal{C}_\mathbb{C}^\infty(S^2), \omega\}$.

Remark 4.1.10. The importance of Hamiltonian vector fields on S^2 is that they generate dynamics of time-dependent smooth functions on the sphere in the same sense of usual Hamilton-Poisson dynamics derived from Newton's laws.

Thus, given a preferred smooth function $h : S^2 \rightarrow \mathbb{R}$, usually called the Hamiltonian function, it generates a flow on the 2-sphere that defines the dynamics of any smooth function $f : S^2 \times \mathbb{R} \rightarrow \mathbb{R}$ via *Hamilton's equation*:

$$\frac{df}{dt} = X_h(f) + \frac{\partial f}{\partial t}, \tag{4.15}$$

which can be rewritten, using (4.8), as

$$\frac{df}{dt} = \{h, f\} + \frac{\partial f}{\partial t}. \tag{4.16}$$

These equations extend naturally to define the dynamics of a complex-valued smooth function $f : S^2 \times \mathbb{R} \to \mathbb{C}$.

Note that the flow of X_h defining the dynamics, denoted $\Phi_h = \{\Phi_h(t), t \in \mathbb{R}\}$, in view of (4.13) satisfies

$$\Phi_h^*(t)\omega = \omega, \forall t \in \mathbb{R}, \tag{4.17}$$

and is therefore a one parameter subgroup of the group of all symplectomorphisms of the 2-sphere, that is, the subgroup of all diffeomorphisms of S^2 preserving the symplectic form ω, which is denoted $Symp(S^2, \omega)$.

Finally, one must remark the close resemblance of Eq. (4.16) to Heisenberg's equation (3.87), a resemblance which is at the core of Bohr's correspondence principle, which will be further explored in Chap. 8.

We recall that Poisson algebras can also be defined on spaces of smooth functions on manifolds of a more general type, called Poisson manifolds. Thus, in the context of G-invariant algebras, $G = SU(2)$, let us now take a closer look at the Poisson algebra of smooth functions on $\mathcal{G}^* \simeq \mathcal{G} = \mathbb{R}^3$ (cf. [48] for more details).

Referring to the extended Hopf map $\mathbb{C}^2 \to \mathbb{R}^3$ given by (2.31), and regarding $\mathbb{R}^3$ as $\mathcal{G}^*$, (x, y, z) can be seen as the "angular momentum" coordinates, so that the Poisson bi-vector field on $\mathcal{G}^*$ is given up to a choice of sign by

$$\Pi = x\partial_y \wedge \partial_z + y\partial_z \wedge \partial_x + z\partial_x \wedge \partial_y$$

and the Poisson bracket of two smooth functions $F, H : \mathcal{G}^* = \mathbb{R}^3 \to \mathbb{C}$ is given by

$$\{F, H\} = \Pi(dF, dH).$$

Now, except at the origin, the foliation of this Poisson manifold $(\mathcal{G}^*, \Pi)$ by symplectic leaves is regular: all symplectic leaves are 2-spheres centered at the origin, which are G-invariant submanifolds of $\mathcal{G}^* = \mathbb{R}^3$ via the coadjoint action.

That is, each of the invariant spheres is a coadjoint orbit of $G = SU(2)$ with its G-invariant symplectic form, defined as in (2.18).

Moreover, from Hamilton's equation (4.16), each of these spheres is also an invariant space under the Poisson dynamics, in other words, the classical dynamics of a $SU(2)$-symmetric mechanical system defined on $(\mathcal{G}^*, \Pi)$ restricts to Poisson dynamics given by (4.16) on each G-invariant symplectic sphere.

From another viewpoint, let us decompose $\mathcal{G}^* - \{0\} = \mathbb{R}^3 - \{0\} \simeq S^2 \times \mathbb{R}^+$ and consider the Poisson algebra $\{C_{\mathbb{C}}^\infty(S^2 \times \mathbb{R}^+), \Pi\}$. It follows that this "extended" Poisson algebra decomposes under the coadjoint action of $SU(2)$ into a (continuous) sum of G-invariant Poisson subalgebras $\{C_{\mathbb{C}}^\infty(S^2), \omega_r\}$, $r \in \mathbb{R}^+$, which are all isomorphic, i.e. $\{C_{\mathbb{C}}^\infty(S^2), \omega_r\} \simeq \{C_{\mathbb{C}}^\infty(S^2), \omega\}$, $\forall r \in \mathbb{R}^+$, under rescaling of ω_r.

Therefore, taking appropriate cautions, as restricting to the space of smooth functions $f : S^2 \times \mathbb{R}^+ \to \mathbb{C}$ with compact support in $\mathbb{R}^+$, etc., we can write:

$$\{C_{\mathbb{C}}^\infty(S^2 \times \mathbb{R}^+), \Pi\} \simeq \{C_{\mathbb{C}}^\infty(S^2), \omega\} \otimes C_{\mathbb{C}}^\infty(\mathbb{R}^+),$$

with $G = SU(2)$ acting trivially on the second factor, that is,

$$G \times C_{\mathbb{C}}^\infty(\mathbb{R}^+) \to C_{\mathbb{C}}^\infty(\mathbb{R}^+), \ (g, f) \mapsto g \cdot f = f, \tag{4.18}$$

and this trivial group action (4.18) on $C_{\mathbb{C}}^\infty(\mathbb{R}^+)$ is the action of every one-parameter group $G = \Phi_H$ obtained by the flow of each specific real Hamiltonian vector field $X_H = \Pi(dH, \cdot)$, $H \in C_{\mathbb{R}}^\infty(S^2 \times \mathbb{R}^+)$, defining such an extended dynamics.

Thus, for a classical spin system, that is, $G = SU(2)$-symmetric classical mechanical system, not much is actually gained by extending the G-invariant Poisson algebra $\{C_{\mathbb{C}}^\infty(S^2), \omega\}$ to the Poisson algebra $\{C_{\mathbb{C}}^\infty(S^2 \times \mathbb{R}^+), \Pi\}$.[1]

On the other hand, as $G = SU(2)$ acts through $SO(3)$ on S^2, this action extends to a G-action on T^*S^2 which is symplectic for the canonical symplectic form $d\eta$ on T^*S^2 and this defines another $G = SU(2)$-invariant Poisson algebra, which can be denoted by $\{C_{\mathbb{C}}^\infty(T^*S^2), d\eta\}$.

But this algebra is "too big" to be the Poisson algebra of the classical spin system because T^*S^2 is a real 4-dimensional symplectic manifold and this corresponds to a classical dynamical system with 2 degrees of freedom (cf. Remark 4.1.2), while quantum spin-j systems are dynamical systems with 1 degree of freedom.[2] Similarly or worse in dimensional counting if we consider $S^2 \times S^2$, T^*S^3, etc... In view of the above discussion, it is natural to make the following definition:

Definition 4.1.11. The *classical spin mechanical system*, or the *classical spin system*, is the homogeneous 2-sphere with its Poisson algebra $\{C_{\mathbb{C}}^\infty(S^2), \omega\}$.

Following standard physics terminology, the homogeneous 2-sphere with its $SO(3)$-invariant area form is called the *phase space* of the classical spin system.

Remark 4.1.12. The homogeneous 2-sphere S^2 with its $SO(3)$-invariant area form ω and metric, given by (4.1) and (4.5) respectively, is also a Kähler (or complex)

---

[1]But more generally, when the foliation of a Poisson manifold by symplectic leaves is irregular, or singular (cf. [82]), the whole Poisson manifold must be used to describe the mechanical system.

[2]Recall that the operator algebra of a spin-j system is generated by the set $\{J_1, J_2, J_3\}$ of three mutually noncommuting Hermitian operators, cf. Eq. (3.61), which furthermore can be reduced to a set of just two noncommuting operators using the commutation relation (3.2), cf. Appendix 10.2 for instance.

manifold and this reflects in the fact that

$$\omega(X_{f_1}, X_{f_2}) = \omega(\nabla f_1, \nabla f_2) \, .$$

We note that the phase space of an affine mechanical system, $(\mathbb{R}^{2k}, \omega)$, can also be seen as a Kähler manifold, $\mathbb{C}^k$. However, this identification depends on the choice of a complex structure and the full group of symmetries of a classical affine mechanical system does not preserve a complex structure.[3] This situation contrasts with the case of the classical spin system: the full symmetry group $SU(2)$, which acts on S^2 via $SO(3)$, preserves the symplectic form (4.1) and the metric (4.5) and therefore it preserves the complex structure that is compatible with both.

At this stage, it is important to emphasize the distinction between the full symmetry group of a mechanical system and its full dynamical group. In the context of the classical spin system, the full symmetry group $SU(2)$, effectively $SO(3)$, is the group of all transformations of the sphere preserving all structures of the classical spin mechanical system, that is, preserving the symplectic form ω *and* the homogeneity of S^2, thus also preserving distances in this case.

On the other hand, the full dynamical group of the classical spin system is the group of all possible Hamilton-Poisson dynamics on S^2. From Remark 4.1.10, each Hamiltonian function defines a specific dynamics, which is a one-parameter subgroup of $Symp(S^2, \omega)$ determined by the flow of its Hamiltonian vector field. Therefore, the full dynamical group of the classical spin system is the group of all such Hamiltonian flows on S^2 starting from the identity, which is denoted $Ham(S^2, \omega)$. From a general perspective, $Ham(S^2, \omega) \subset Symp_0(S^2, \omega) \subset Symp(S^2, \omega)$, where $Symp_0(S^2, \omega)$ is the connected component of $Symp(S^2, \omega)$. But for S^2 we have equality, $Ham(S^2, \omega) = Symp_0(S^2, \omega) = Symp(S^2, \omega)$.[4]

The Lie algebra of $Ham(S^2, \omega)$ is the Lie algebra of real Hamiltonian vector fields, which from (4.14) is (modulo constant functions) isomorphic to the Poisson algebra of real smooth functions on S^2. And as this Poisson algebra is infinite dimensional, so is $Ham(S^2, \omega) = Symp(S^2, \omega)$. Therefore, we see that the full symmetry group is a very small subgroup of the full dynamical group.

Remark 4.1.13. The same distinction applies to a quantum spin-j mechanical system. Its full dynamical group is the group of all possible Heisenberg dynamics on $\mathcal{H}_j$. From Remark 3.3.13, each Hermitian matrix generates a one-parameter subgroup of $U(n + 1)$ which defines a specific dynamics on $\mathcal{H}_j$. Therefore, the symmetry group $SU(2)$ is a proper subgroup of the full dynamical group of the spin-j system: the connected component of $U(n + 1)$, which is the whole $U(n + 1)$.

---

[3] We refer to the next chapter, Intermission, for the definition of a classical affine mechanical system $(\mathbb{R}^{2k}, \omega)$ with its symplectic form ω, as well as its symmetry group, etc.

[4] The first equality follows from S^2 being a simply connected compact manifold and the second from $Symp(S^2, \omega)$ being homotopic to $SO(3)$, which is path-connected. For general symplectic manifolds, these inclusions are often strict and not always well understood. We refer the interested reader to some texts in symplectic topology, e.g. [49, 54].

4.2   SO(3)-Invariant Decomposition of the Poisson Algebra

Having defined the Poisson algebra of the classical spin system, we now study how it decomposes under the action of $SO(3)$, as this will be fundamental for what will follow later on. To this end, we must look at the polynomial algebra on S^2.

4.2.1   The Irreducible Summands of the Polynomial Algebra

Let $\mathbb{R}[x, y, z]$ be the algebra of real polynomials on $\mathbb{R}^3$. Their restriction to S^2 defines a distinguished class of functions densely approximating smooth functions,

$$\mathbb{R}[x, y, z] \to \mathbb{R}[x, y, z] / \langle x^2 + y^2 + z^2 - 1 \rangle \simeq Poly_{\mathbb{R}}(S^2) \subset C_{\mathbb{R}}^{\infty}(S^2). \qquad (4.19)$$

We regard these spaces as the real form of the spaces of $\mathbb{C}$-valued functions, that is, of their complexified versions

$$\mathbb{C}[x, y, z] / \langle x^2 + y^2 + z^2 - 1 \rangle \simeq Poly_{\mathbb{C}}(S^2) \subset C_{\mathbb{C}}^{\infty}(S^2). \qquad (4.20)$$

These function spaces are $SO(3)$-modules with the induced action

$$F \to F^g, \quad F^g(\mathbf{n}) = F(g^{-1}\mathbf{n}) \quad . \qquad (4.21)$$

In particular, $g \in SO(3)$ transforms a polynomial Y by substituting the variables

$$Y(x, y, z) \to Y(x', y', z') = Y^g(x, y, z)$$

where $(x', y', z') = (x, y, z)g$ (matrix product).

Remark 4.2.1. Fixing a positive orientation for $\{x, y, z\}$, we adopt the universal convention of identifying the positive z direction with the "north pole" $\mathbf{n}_0 \in S^2$.

Then with the basis $\{x, y, z\}$, linear forms transform by $g \in SO(3)$ according to the standard representation ψ_1 on $\mathbb{R}^3$, that is, $\psi_1(g) = g$, and the action on forms of degree l is the l-th symmetric tensor product of ψ_1, with the splitting

$$\mathbb{R}[x, y, z]_l : S^l \psi_1 = \psi_l + \psi_{l-2} + \psi_{l-4} + \ldots .$$

On the other hand, multiplication by $(x^2 + y^2 + z^2)$ is injective in the polynomial ring, and clearly the lower summands lie in the subspace

$$\mathbb{R}[x, y, z]_{l-2} (x^2 + y^2 + z^2) \subset \mathbb{R}[x, y, z]_l : \psi_{l-2} + \psi_{l-4} + \ldots .$$

Consequently, when we restrict functions to S^2 we identify them according to (4.19), and then there will be a unique irreducible summand of type ψ_l for each l, spanned by polynomials of proper (minimal) degree l. In this way, we have obtained the following result:

Proposition 4.2.2. *Real (resp. complex) polynomial functions of proper degree $\leq n$ on S^2 constitute a $SO(3)$-representation with the same splitting into irreducibles as the space of Hermitian matrices (resp. the full matrix space) in dimension $n + 1$:*

$$Poly_{\mathbb{R}}(S^2)_{\leq n} = \sum_{l=0}^{n} Poly(\psi_l) \simeq \sum_{l=0}^{n} \mathcal{H}(\psi_l) = \mathcal{H}(n + 1), \quad cf.\,(3.58), \qquad (4.22)$$

$$Poly_{\mathbb{C}}(S^2)_{\leq n} = \sum_{l=0}^{n} Poly(\varphi_l) \simeq \sum_{l=0}^{n} M_{\mathbb{C}}(\varphi_l) = M_{\mathbb{C}}(n + 1), \quad cf.\,(3.56). \qquad (4.23)$$

Definition 4.2.3. $Poly(\psi_l)$ (resp. its complex extension $Poly(\varphi_l)$) denotes the space of *spherical harmonics* of type ψ_l (resp. φ_l), $0 \leq l \leq n$. In view of the relation $x^2 + y^2 + z^2 = 1$ these are polynomial functions of proper degree l.

4.2.2 The Standard Basis of Spherical Harmonics

For the sake of completeness, let us also give a procedure for the calculation of spherical harmonics, namely polynomials of proper degree l which constitute a standard orthonormal basis

$$Y_{l,l}, Y_{l,l-1}, \ldots, Y_{l,0}, \ldots, Y_{l,-l+1}, Y_{l,-l} \qquad (4.24)$$

for the representation space $Poly(\varphi_l) \simeq \mathbb{C}^{2l+1}$, $l \geq 1$. Sometimes we also use the notation Y_l^m for $Y_{l,m}$.

The basic case is $l = 1$, where $g \in SO(3)$ transforms linear forms, expressed in the basis $\{x, y, z\}$, by the same matrix g. Namely, the infinitesimal generators $L_k \in \mathcal{SO}(3)$ are the matrices in (2.20), and the angular momentum operators act linearly on $\mathbb{C}\{x, y, z\} \simeq \mathbb{C}^3$ with the matrix representation

$$J_k = iL_k \,, k = 1, 2, 3 \,, \ J_{\pm} = J_1 \pm iJ_2 \,,$$

for L_k given by (2.20). Then, for example, $J_3(x) = iy$, $J_3(y) = -ix$. The operators J_k and $J_{\pm}$ act as derivations on functions in general, for example

$$J_3((x + iy)^l) = l(x + iy)^l$$

and consequently $Y_{l,l}$ must be proportional to $(x + iy)^l$.

Remark 4.2.4. As in the definition of the standard coupled basis of operators (cf. Sect. 3.3.2), the definition of a standard orthonormal basis for each $Poly(\varphi_l) \simeq \mathbb{C}^{2l+1}$ depends on a choice of overall phase (one for each $Poly(\varphi_l)$), see Remark 3.2.5 and Proposition 3.3.5.

We set $Y_{0,0} = 1$ and choose the phase convention by setting

$$Y_{l,l} = \frac{(-1)^l}{\lambda_{l,l}} (x + iy)^l, \ \forall l \in \mathbb{N}, \ Y_{l,m-1} = \frac{1}{\beta_{l,m}} J_-(Y_{l,m}), \ \text{for } 0 < |m| < l,$$

$$(4.25)$$

where $\lambda_l = \lambda_{l,l} > 0$ is calculated by

$$\lambda_l^2 = \| (x + iy)^l \|^2 = \frac{1}{4\pi} \int_{S^2} (1 - z^2)^l dS = \sum_{k=0}^{l} \binom{l}{k} \frac{(-1)^k}{2k+1} = \frac{(l!)^2 2^{2l}}{(2l+1)!}.$$

$$(4.26)$$

With reference to (3.17), (3.18), (3.77) and (6.17), we also conclude

$$Y_{l,m} = \frac{(-1)^l}{\lambda_{l,m}} J_-^{l-m} (x + iy)^l, \quad \lambda_{l,m} = \frac{l! \, 2^l}{\sqrt{2l+1}} \sqrt{\frac{(l-m)!}{(l+m)!}}, \ 0 \leq m \leq l,$$

$$(4.27)$$

and changing $m \to -m$ has the effect

$$Y_{l,-m} = (-1)^m \overline{Y_{l,m}}, \ 0 \leq m \leq l. \tag{4.28}$$

Finally, we have the following relations, that set $Y_{l,m}$ as eigenvector of J_3 and J^2:

$$J_3 Y_{l,m} = m Y_{l,m}, \ \sum_{k=1}^{3} J_k^2 Y_{l,m} = l(l+1) Y_{l,m}. \tag{4.29}$$

It should be mentioned that $Y_{l,0}$ depends only on z. For easy reference we list the resulting functions for $l = 1, 2, 3$:

$$Y_{1,1} = -\sqrt{\frac{3}{2}}(x + iy), \ Y_{1,0} = \sqrt{3}z, \ Y_{1,-1} = \sqrt{\frac{3}{2}}(x - iy),$$

$$Y_{2,2} = \sqrt{\frac{15}{8}}(x + iy)^2, \ Y_{2,1} = -\sqrt{\frac{15}{2}}(x + iy)z, \ Y_{2,0} = \frac{\sqrt{5}}{2}(3z^2 - 1), \quad (4.30)$$

$$Y_{3,3} = -\frac{\sqrt{35}}{4}(x + iy)^3, \ Y_{3,2} = \frac{3}{2}\sqrt{\frac{35}{6}}(x + iy)^2 z,$$

$$Y_{3,1} = -\frac{\sqrt{21}}{4}(x + iy)(5z^2 - 1), \ Y_{3,0} = \frac{\sqrt{7}}{2}(5z^3 - 3z).$$

Remark 4.2.5. As exemplified above, we recall that all spherical harmonics may be homogenized by using the relation $x^2 + y^2 + z^2 = 1$. Thus, if $\mathbf{n} \in S^2$ has cartesian coordinates (x, y, z), so that $Y_{l,m} \equiv Y_{l,m}(x, y, z) = Y_{l,m}(\mathbf{n})$, and $-\mathbf{n} \in S^2$ denotes the antipodal point to $\mathbf{n}$, with coordinates $(-x, -y, -z)$, then

$$Y_{l,m}(-\mathbf{n}) = (-1)^l \, Y_{l,m}(\mathbf{n}). \tag{4.31}$$

Because it is not always so easy to determine the proper degree of a spherical polynomial expressed as a polynomial in the cartesian coordinates (x, y, z), in view of the relation $x^2 + y^2 + z^2 = 1$, it is useful to have formulae for the spherical harmonics which are expressed in terms of spherical coordinates (φ, θ), cf. (2.39).

These are well known and, with our previous scaling convention, we have

$$Y_{l,m} = Y_l^m = \sqrt{2l + 1}\sqrt{\frac{(l - m)!}{(l + m)!}} \, P_l^m(\cos\varphi)e^{im\theta}, \tag{4.32}$$

$$Y_{l,m}(0, \theta) = \delta_{m,0}\sqrt{2l + 1} \tag{4.33}$$

where the functions P_l^m are the so-called associated Legendre polynomials, which are "classical" well-known polynomials in $\cos\varphi = z$ and $\sin\varphi = (1 - z^2)^{1/2}$ and the identity (4.33) simply means that $Y_{l,m}$ vanishes at the north pole, except when $m = 0$, in which case the value is $\sqrt{2l + 1}$.

More precisely, the functions $P_l = P_l^0$ are the Legendre polynomials, normalized so that $P_l(1) = 1$. As polynomials in z, they are defined by

$$P_l(z) = \frac{1}{2^l}\sum_{k=0}^{[l/2]}(-1)^k\frac{(2l - 2k)! \, z^{l-2k}}{k!(l - k)!(l - 2k)!}, \tag{4.34}$$

where $[q]$ denotes the integral part of $q \in \mathbb{Q}$, or by Rodrigues' formula:

$$P_l(z) = \frac{1}{2^l l!}\frac{d^l}{dz^l}(z^2 - 1)^l. \tag{4.35}$$

They can also be written as polynomials in $(1 + z)$ by

$$P_l(z) = \sum_{k=0}^{l}(-1)^{l+k}\binom{l + k}{k}\binom{l}{k}\left(\frac{1 + z}{2}\right)^k = \sum_{k=0}^{l}\frac{(-1)^{l+k}}{(k!)^2}\frac{(l + k)!}{(l - k)!}\left(\frac{1 + z}{2}\right)^k \tag{4.36}$$

and they satisfy the following differential equation:

$$\frac{d}{dz} P_{l+1}(z) = (2l + 1) P_l(z) + (2(l - 2) + 1) P_{l-2}(z) + (2(l - 4) + 1) P_{l-4}(z) + \cdots$$
(4.37)

$$\Longleftrightarrow \quad (2l + 1) P_l(z) = \frac{d}{dz}[P_{l+1}(z) - P_{l-1}(z)] .$$
(4.38)

Furthermore, the Legendre polynomials are orthogonal in the interval $(-1, 1)$

$$\int_{-1}^{1} P_l(z) P_k(z) dz = \delta_{l,k} \frac{2}{2l + 1} .$$
(4.39)

The associated Legendre polynomials P_l^m are defined for $m \geq 0$ by

$$P_l^m(z) = (-1)^m (1 - z^2)^{m/2} \frac{d^m}{dz^m} P_l(z) , \quad m \geq 0$$
(4.40)

and for negative m by the identity

$$P_l^{-m} = (-1)^m \frac{(l - m)!}{(l + m)!} P_l^m .$$
(4.41)

By setting $P_0^0 = 1$, $P_l^m = 0$ for $l < m$, and

$$P_l^l(z) = (-1)^l (2l - 1)!!(1 - z^2)^{l/2},$$
(4.42)

the polynomials P_l^m can also be derived by the recurrence formula

$$(l - m) P_l^m(z) = (2l - 1)z P_{l-1}^m(z) - (l + m - 1) P_{l-2}^m(z) .$$
(4.43)

The first polynomials are given by:

$$P_1^0 = z, \; P_1^1 = -(1 - z^2)^{1/2}, \; P_2^0 = \frac{1}{2}(3z^2 - 1), \; P_2^1 = -3z(1 - z^2)^{1/2}, \; P_2^2 = 3(1 - z^2),$$

$$P_3^0 = \frac{1}{2}z(5z^2 - 3), \; P_3^1 = \frac{3}{2}(1 - 5z^2)(1 - z^2)^{1/2}, \; P_3^2 = 15z(1 - z^2), \; P_3^3 = -15(1 - z^2)^{3/2}$$

As mentioned before, the Legendre and associated Legendre polynomials are "classical" very well-known polynomials. Our concise presentation above is meant for partial self-containment of the monograph and the interested reader can find ample material on these polynomials in various books, handbooks and websites.

4.2.3 Decompositions of the Classical Products

The space of complex polynomials on the 2-sphere defined by (4.20), $Poly_{\mathbb{C}}(S^2)$, densely approximates the space of smooth functions on the 2-sphere, $C_{\mathbb{C}}^{\infty}(S^2)$, and therefore it densely approximates the Poisson algebra of the classical spin system, $\{C_{\mathbb{C}}^{\infty}(S^2), \omega\}$, cf. Definitions 4.1.9 and 4.1.11, by letting $n \to \infty$ for

$$Poly_{\mathbb{C}}(S^2)_{\leq n} \subset C_{\mathbb{C}}^{\infty}(S^2) .$$

Thus, although the classical products of functions on the sphere, the pointwise product and the Poisson bracket, are defined for general smooth functions on S^2, for what follows it is useful to have formulas for these classical products as decomposed in the standard orthonormal basis of spherical harmonics $Y_{l,m}$ satisfying

$$\langle Y_{l,m}, Y_{l',m'} \rangle = \delta_{l,l'} \delta_{m,m'} .$$

The formula for the pointwise product has long been known, although it is not so commonly presented with its proof. Here we prove it following the approach outlined in [63], which uses a connection between the above functions and the Wigner D-functions. On the other hand, as far as we know, the formula for the Poisson bracket appeared for the first time only in 2002, in a paper by Freidel and Krasnov [30], on which we based our proof (further below, in Appendix 10.4).

Decomposition of the Pointwise Product

Proposition 4.2.6. *The pointwise product of spherical harmonics decomposes in the basis of spherical harmonics according to the following formula:*

$$Y_{l_1,m_1} Y_{l_2,m_2} = \sum_{\substack{l=|l_1-l_2| \\ l \equiv l_1+l_2 \,(\mathrm{mod}\, 2)}}^{l_1+l_2} \sqrt{\frac{(2l_1+1)(2l_2+1)}{2l+1}} C_{m_1,m_2,m}^{l_1,l_2,l} C_{0,0,0}^{l_1,l_2,l} Y_{l,m} \qquad (4.44)$$

Proof. The elements $g \in SO(3)$ rotate the points p in the Euclidean 3-space and its unit sphere S^2, but fixing the points we can also view a rotation as a transformation of the coordinate system (x, y, z) to another system (x', y', z'). Then a function Y on the sphere is transformed to a function Y^g such that

$$Y^g(x, y, z) = Y(x', y', z'), \ Y^g(\varphi, \theta) = Y(\varphi', \theta'),$$

and on the other hand

$$Y_{l,m}^g = \sum_{\mu} D_{\mu,m}^l(g) Y_{l,\mu} \qquad (4.45)$$

since $g \to D^l(g)$ is, by definition, the matrix representation of $SO(3)$ on the space spanned by the functions $\{Y_{l,m}\}$. Now, for a fixed positive integer l and two points $p_1 = (\varphi_1, \theta_1)$, $p_2 = (\varphi_2, \theta_2)$ on the sphere, we claim that the quantity

$$\Theta = \sum_m \overline{Y}_{l,m}(\varphi_1, \theta_1) Y_{l,m}(\varphi_2, \theta_2) \tag{4.46}$$

is invariant under rotations. In fact, by (4.45) and the unitary property of the matrix $D^l = D^l(g)$,

$$\Theta^g = \sum_m \sum_{\mu_1, \mu_2} \overline{D}^l_{\mu_1, m} D^l_{\mu_2, m} \, \overline{Y}_{l,\mu_1}(\theta_1, \varphi_1) Y_{l,\mu_2}(\varphi_2, \theta_2)$$

$$= \sum_{\mu_1, \mu_2} \delta_{\mu_1, \mu_2} \, \overline{Y}_{l,\mu_1}(\varphi_1, \theta_1) Y_{l,\mu_2}(\varphi_2, \theta_2) = \sum_\mu \overline{Y}_{l,\mu}(\varphi_1, \theta_1) Y_{l,\mu}(\varphi_2, \theta_2) = \Theta$$

In particular, let φ be the spherical distance between p_1 and p_2, and choose g to rotate the coordinate system so that $\varphi_1' = 0$ and $\theta_2' = 0$. Then p_1 is the new north pole and hence $\varphi_2' = \varphi$, and by (4.33) the quantity (4.46) becomes

$$\Theta = \overline{Y}_{l,0}(0, \theta_1') Y_{l,0}(\varphi, 0) = \sqrt{2l+1} Y_{l,0}(\varphi, 0).$$

Combined with the first formula (4.46) of Θ this yields the general formula

$$Y_{l,0}(\varphi, 0) = \frac{1}{\sqrt{2l+1}} \sum_m \overline{Y}_{l,m}(\varphi_1, \theta_1) Y_{l,m}(\varphi_2, \theta_2) : \tag{4.47}$$

On the other hand, let us evaluate the identity (4.45) with $m = 0$ at the point p_2 and obtain the following identity similar to (4.47):

$$Y_{l,0}(\varphi_2', 0) = Y_{l,0}(\varphi_2', \theta_2') = \sum_m D^l_{m,0}(g) Y_{l,m}(\varphi_2, \theta_2). \tag{4.48}$$

Now, choose the rotation $g = R(\alpha, \beta, 0)$ with Euler angle $\gamma = 0$ (cf. (2.26)), which rotates the (old) north pole $(0, 0, 1)$ to the point

$$p_1 = (x_1, y_1, z_1) = (\cos\alpha \sin\beta, \sin\alpha \sin\beta, \cos\beta),$$

namely the new north pole is $p_1 = (\varphi_1, \theta_1) = (\beta, \alpha)$ and therefore its spherical distance to p_2 is φ_2'. Consequently, the left sides of (4.47) and (4.48) are identical, both expressing the same general coupling rule, so we conclude

$$Y_{l,m}(\beta, \alpha) = \sqrt{2l+1} \overline{D}^l_{m,0}(\alpha, \beta, 0).$$

Combining the above identity with the coupling rule (3.45) for D-functions, we deduce the formula (4.44) as follows:

$$Y_{l_1,m_1}(\varphi,\theta)Y_{l_2,m_2}(\varphi,\theta) = \sqrt{(2l_1+1)(2l_2+1)}\overline{D}^{l_1}_{m_1,0}(\theta,\varphi,0)\overline{D}^{l_2}_{m_2,0}(\theta,\varphi,0)$$

$$= \sqrt{(2l_1+1)(2l_2+1)}\sum_l C^{l_1,l_2,l}_{m_1,m_2,m}C^{l_1,l_2,l}_{0,0,0}\overline{D}^l_{m_1+m_2,0}(\theta,\varphi,0)$$

$$= \sum_{l=|l_1-l_2|}^{l_1+l_2}\sqrt{\frac{(2l_1+1)(2l_2+1)}{2l+1}}C^{l_1,l_2,l}_{m_1,m_2,m}C^{l_1,l_2,l}_{0,0,0}Y_{l,m}(\varphi,\theta)$$

In the above sum the range of l is also restricted to $l \equiv l_1 + l_2 (\mathrm{mod}\,2)$, since the coefficient $C^{l_1,l_2,l}_{0,0,0}$ vanishes when $l + l_1 + l_2$ is odd. $\qquad\square$

Remark 4.2.7. We note that the particular rightmost Clebsch-Gordan coefficient appearing in Eq. (4.44) has a simple closed formula (cf. [74]):

$$C^{l_1,l_2,l_3}_{0,0,0} = \frac{(-1)^{(l_1+l_2-l_3)/2}\sqrt{2l_3+1}\Delta(l_1,l_2,l_3)((l_1+l_2+l_3)/2)!}{((-l_1+l_2+l_3)/2)!((l_1-l_2+l_3)/2)!((l_1+l_2-l_3)/2)!}, \qquad (4.49)$$

for $L = l_1 + l_2 + l_3$ even, and $C^{l_1,l_2,l_3}_{0,0,0} = 0$ for L odd, where $\Delta(l_1,l_2,l_3)$ is given by Eq. (3.48). We also note that the vanishing condition above follows from one of the symmetries of the Clebsch-Gordan coefficients (cf. (3.50)), namely,

$$C^{l_1,l_2,l_3}_{0,0,0} = (-1)^{l_1+l_2+l_3}C^{l_1,l_2,l_3}_{-0,-0,-0}. \qquad (4.50)$$

Decomposition of the Poisson Bracket

Proposition 4.2.8 ([30]). *The Poisson bracket of spherical harmonics decomposes in the basis of spherical harmonics according to the following formula:*

$$i\left\{Y^{m_1}_{l_1},Y^{m_2}_{l_2}\right\} = \sum_{\substack{l=|l_1-l_2|+1 \\ l\equiv l_1+l_2-1}}^{l_1+l_2-1}\sqrt{\frac{(2l_1+1)(2l_2+1)}{2l+1}}C^{l_1,\,l_2,\,l}_{m_1,m_2,m}P(l_1,l_2,l)\,Y^m_l$$

$$(4.51)$$

where, by definition,

$$P(l_1,l_2,l_3) \qquad\qquad\qquad\qquad\qquad\qquad\qquad\qquad\qquad (4.52)$$

$$= \frac{(-1)^{(l_1+l_2-l_3+1)/2}\sqrt{2l_3+1}\Delta(l_1,l_2,l_3)(L+1)((L-1)/2)!}{((-l_1+l_2+l_3-1)/2)!((l_1-l_2+l_3-1)/2)!((l_1+l_2-l_3-1)/2)!},$$

for $L = l_1 + l_2 + l_3$ odd, and $P(l_1,l_2,l_3) = 0$ for L even.

The proof of Proposition 4.2.8 for the decomposition of the Poisson bracket is a bit more technical and considerably longer than the proof of the decomposition of the pointwise product, so it has been placed in Appendix 10.4.

Remark 4.2.9. One should note the close resemblance between formula (4.44) for the pointwise product and formula (4.51) for the Poisson bracket, in view of the close resemblance between formulas (4.49) and (4.52) for the coefficients (which depend only on the l's) multiplying $C_{m_1,m_2,m}^{l_1,l_2,l}$ in each case.

Note also that, since the Poisson bracket is skew symmetric, the summation in (4.51) starts with $l = |l_2 - l_1| + 1 = \max\{ |l_1 - l_2 - 1|, |l_2 - l_1 - 1| \}$.

Remark 4.2.10. Of course, by linearity, for f and g decomposed in the basis of spherical harmonics, one obtains the coefficients of the expansion in spherical harmonics of fg and $\{f, g\}$ straightforwardly from (4.44)–(4.49) and (4.51)–(4.52).

Chapter 5
Intermission

Brief historical overview of symbol correspondences in affine mechanical systems The names of Pythagoras, Euclid and Plato[1] can perhaps best summarize the dawning of the "mathematization of nature" process, that took place in ancient Greek civilization when the Pythagorean school boosted the philosophy that numbers and mathematical concepts were the key to understand the divine cosmic order. While Euclid's *Elements* set forth the axiomatization of geometry, what we now call "flat" geometry, it was Plato, however, who first applied the Pythagorean philosophy to the empirical science of the time: astronomy.

Inspired by the "divine geometrical perfection" of spheres and circles, in his dialogue *Timaeus* Plato set forth the notion of uniform circular motions to be the natural motions of heavenly bodies, as each of which being in possession of its own anima mundi, or "world soul", would continue on such an eternal motion.

Plato's "circular inertial motions" had to come to terms with the empirical astronomical data of his historical period, however, and thus a whole mathematical model was developed, based on uniform circular motions, in which circles upon circles upon circles ... were used to describe the apparent motions of the sun, moon and known planets, culminating in the treatise of Ptolemy called the *Almagest*. The latter was used by professional astronomers for centuries with adequate precision, as it realized the first "perturbation theory" to be used in history, wherein new tinier circles could always be added to better fit new and more precise data, in a process akin to our well-known Fourier series decomposition of periodic functions.

Many centuries later, when Galileo discovered empirically that a principle of inertial motions applied to earthly motions as well, he then followed Plato's principle, so that Galileo's inertial motions were still the circular inertial motions of Plato, only extended to sub-celestial motions (after all, a uniform straight motion on the surface of the earth is actually circular) [43].

---

[1]This list is by no means exclusive and other names could/should be added, like for instance, Eudoxus, Archimedes, Apollonius...

© Springer International Publishing Switzerland 2014
P. de M. Rios, E. Straume, *Symbol Correspondences for Spin Systems*,
DOI 10.1007/978-3-319-08198-4_5

It was Descartes who reformulated the universal principle of inertia in terms of uniform linear motions. But then, the approximately circular planetary motions (which at that time had already been found by Kepler to be much more simply described/approximated by ellipses) had to be re-explained as resulting from the action of a "force", as emphasized by Hooke, culminating in Newton's mechanical theory of universal gravitation set forth in his *Principia*.

Thus, following Descartes, Newton placed uniform linear motion, or the straight line, as the first mathematical axiom of his new mechanics, greatly increasing in importance Euclid's axiomatics of geometry, so that Euclidean 3-space became synonymous with universal space.

And as Newtonian mechanics developed mathematically throughout decades, straight lines retained their primordial predominance so that, at the turn of the twentieth century, when Planck first set forth the hypothesis of quantized energies, the mechanics of conservative systems was modeled on what we now understand as the Poisson algebra of functions on a symplectic affine space, which doubled the dimensions of the configuration space of positions. This latter was seen as some product of Euclidean 3-spaces, or in simplified versions as an Euclidean k-space $\mathbb{R}_\epsilon^k$, including $k = 1$, a straight line (in our notation, $\mathbb{R}^k$ is the kth power of $\mathbb{R}$, while $\mathbb{R}_\epsilon^k$ also carries the Euclidean metric ϵ_k).

In this way, the most fundamental symmetry of these spaces are straight linear motions, also referred to as rectilinear motions, or affine translations, and so we can call mechanical systems which are symmetric under such affine translations as *affine mechanical systems*.

For an affine symplectic space $(\mathbb{R}^{2k}, \omega) \equiv \mathbb{R}_\omega^{2k}$, where $\mathbb{R}^{2k} \ni (p, q)$, $p, q \in \mathbb{R}^k$,

$$\omega = \sum_{i=1}^{k} dp_i \wedge dq_i \,,$$

the group of affine translations is $\mathbb{R}^{2k}$ with usual vector addition as the group product, and such that for any fixed $\xi = (a, b) \in \mathbb{R}^{2k}$ the affine translation is

$$T(\xi) : \mathbb{R}_\omega^{2k} \to \mathbb{R}_\omega^{2k} \,, \ (p, q) \mapsto (p + a, q + b) \,. \tag{5.1}$$

In fact, it is well known that the full group of symmetries of an affine symplectic space $\mathbb{R}_\omega^{2k}$ is actually much larger and is called the affine symplectic group, $aSp_{\mathbb{R}}(2k) = \mathbb{R}^{2k} \rtimes Sp_{\mathbb{R}}(2k)$, the semi-direct product of $\mathbb{R}^{2k}$ and $Sp_{\mathbb{R}}(2k)$, the latter being the group of linear symplectic transformations of $\mathbb{R}_\omega^{2k}$, which is the maximal subgroup of $GL_{\mathbb{R}}(2k)$ that preserves the symplectic structure ω.

Therefore, when a mathematical formulation of quantum mechanical systems began taking shape in early twentieth century, it also followed the form of an affine mechanical system. However, in contrast to classical affine mechanical systems, which are defined by the Poisson algebra of functions on a finite dimensional affine symplectic space, quantum affine mechanical systems are defined by the algebra of operators on an infinite dimensional Hilbert space, $\mathcal{H} = L_{\mathbb{C}}^2(\mathbb{R}_\epsilon^k)$, where $\mathbb{R}_\epsilon^k$ is usually identified with either the space of positions q or the space of momenta p.

Because, while quantum symmetries have to be implemented by unitary operators, according to the mathematical framework established by von Neumann, some of these arise from the symmetries of $\mathbb{R}_\epsilon^k$ itself, whose symmetry group is the Euclidean group $E_{\mathbb{R}}(k) = \mathbb{R}^k \rtimes O(k)$, the semi-direct product of $\mathbb{R}^k$ and the orthogonal group $O(k)$, the latter being the maximal subgroup of $GL_{\mathbb{R}}(k)$ that preserves distances (and hence also angles).

Now, $O(k)$ is compact and thus admits faithful finite dimensional unitary representations, but $\mathbb{R}^k$ and hence $E_{\mathbb{R}}(k)$ are noncompact, so that both only admit almost faithful unitary representations that are infinite dimensional. This is the geometrical explanation of why the Hilbert spaces of quantum affine mechanical systems have to be infinite dimensional.

Letting $E_{\mathbb{R}}(k)$ act "in the same way", or diagonally, on the p and q subspaces of $\mathbb{R}_\omega^{2k}$, $E_{\mathbb{R}}(k)$ is naturally a subgroup of $aSp_{\mathbb{R}}(2k)$, the latter group being much larger. Let us denote this embedding of $E_{\mathbb{R}}(k)$ into $aSp_{\mathbb{R}}(2k)$ by $\tilde{E}_{\mathbb{R}}(k)$. Then,

$$\tilde{E}_{\mathbb{R}}(k) \subset E_{\mathbb{R}}(k \oplus k), \ \mathbb{R}^{2k} \subset E_{\mathbb{R}}(k \oplus k),$$

$$E_{\mathbb{R}}(k \oplus k) \subset E_{\mathbb{C}}(k) \subset aSp_{\mathbb{R}}(2k),$$

where $E_{\mathbb{R}}(k \oplus k) = \mathbb{R}^{2k} \rtimes \tilde{O}(k)$, with $\tilde{O}(k)$ denoting the embedding of $O(k)$ into $Sp_{\mathbb{R}}(2k)$ obtained by diagonal action on the p and q subspaces, and where $E_{\mathbb{C}}(k) = \mathbb{C}^k \rtimes U(k)$ is the "complex Euclidean group", which is the subgroup of $aSp_{\mathbb{R}}(2k)$ that preserves a complex structure on $\mathbb{R}^{2k} \simeq \mathbb{C}^k$, with $U(k) \simeq Sp_{\mathbb{R}}(2k) \cap O(2k)$.

By first restricting attention to affine translations and noting that the group of translations of $\mathbb{R}_\omega^{2k}$ has the double dimension of $\mathbb{R}^k$, a natural question was how to extend the unitary action of $\mathbb{R}^k$ on $\mathcal{H}$ to a unitary action of $\mathbb{R}^{2k}$ in a way to account for Heisenberg's canonical commutation relations $[\hat{p}_i, \hat{q}_j] = i\hbar\delta_{ij}$.

A nice approach to solving this problem is by presenting the Heisenberg group H_0^{2k}, which is a $U(1)$ central extension of $\mathbb{R}^{2k}$, with product

$$(\xi_1, \exp(i\theta_1)) \cdot (\xi_2, \exp(i\theta_2)) = (\xi_1 + \xi_2, \exp(i(\theta_1 + \theta_2 + (\omega(\xi_1, \xi_2)/2\hbar)))) , \tag{5.2}$$

where $\xi_j \in \mathbb{R}^{2k}$, $\exp(i\theta_j) \in U(1)$, $j = 1, 2$. Working with a unitary action of H_0^{2k} on $\mathcal{H}$, and the Fourier transform, we can formally set up Weyl's symbol correspondence between operators on $\mathcal{H}$ and functions on $\mathbb{R}_\omega^{2k}$, the latter "almost identified" with H_0^{2k}, as the commutator of (5.2) does not depend on phases θ_j.

More concretely, let $\xi = (a, b) \in \mathbb{R}_\omega^{2k}$, as in (5.1), and recall that ϵ_k denotes the Euclidean metric on $\mathbb{R}_\epsilon^k$, in other words, the usual scalar product. Defining

$$T_0^\hbar(\xi, \exp(i\theta)) : \mathcal{H} \to \mathcal{H} ,$$

$$\psi(q) \mapsto \psi(q - b) \exp(i\theta + \epsilon_k(a, q)/\hbar - \epsilon_k(a, b)/2\hbar) , \tag{5.3}$$

one can verify that $T_0^\hbar(\xi, \exp(i\theta))$ is an element in a unitary representation of H_0^{2k}, acting on $\mathcal{H}$ (see e.g. [45]). Choosing $T_0^\hbar(\xi, 1)$ to represent the equivalence class $[T_0^\hbar(\xi)]$ defined by the equivalence relation $T_0^\hbar(\xi, \exp(i\theta_1)) \approx T_0^\hbar(\xi, \exp(i\theta_2))$, then via Fourier transform we can formally define the "reflection" operator on $\mathcal{H}$ by

$$R_0^\hbar(x) = \frac{1}{(2\pi\hbar)^k} \int T_0^\hbar(\xi, 1) \exp(i\omega(x, \xi)/\hbar)d\xi \; : \; \mathcal{H} \to \mathcal{H}, \tag{5.4}$$

where $d\xi$ is the Liouville volume element on $\mathbb{R}_\omega^{2k} \ni x, \xi$.

Then, if A is an operator on $\mathcal{H}$, the Weyl symbol of A is the complex function $W_A^\hbar$ on $\mathbb{R}_\omega^{2k}$ formally defined by (see [64], also [51]):

$$W_A^\hbar(x) = trace(AR_0^\hbar(x)), \tag{5.5}$$

in such a way that, if the action of A on $\mathcal{H}$ is described in integral form by

$$A\psi(q') = \int S_A^\hbar(q', q'')\psi(q'')d^k q'', \tag{5.6}$$

then, writing $x = (p, q)$, the integral kernel $S_A^\hbar$ and the symbol $W_A^\hbar$ are related by

$$W_A^\hbar(p, q) = \frac{1}{(2\pi\hbar)^k} \int S_A^\hbar(q - v/2, q + v/2) \exp(i\epsilon_k(v, p)/\hbar)d^k v, \tag{5.7}$$

$$S_A^\hbar(q', q'') = \frac{1}{(2\pi\hbar)^k} \int W_A^\hbar(p, (q' + q'')/2) \exp(i\epsilon_k(p, q' - q'')/\hbar)d^k p, \tag{5.8}$$

where $d^k q'', d^k v, d^k p$ denote the usual Euclidean volume $d^k p = dp_1 dp_2 \cdots dp_k$.

Whenever well defined, Weyl's correspondence, seen as a map $A \mapsto W_A^\hbar$ that assigns to an operator A a unique function $W_A^\hbar$ on affine symplectic space, satisfies:

(i)   It is a linear injective map;

(ii) $W_{A^*}^\hbar = \overline{W_A^\hbar}$, where A^* is the adjoint of A;

(iii) It is equivariant under action of $aSp_\mathbb{R}(2k)$,

where the action on functions is the usual one, while the action on operators is the effective action of $aSp_\mathbb{R}(2k)$ obtained via adjoint action of a (∞-dimensional) unitary representation of the affine metaplectic group, the latter being a $U(1)$ central extension of $aSp_\mathbb{R}(2k)$ containing as subgroups H_0^k and the metaplectic group $Mp(2k)$, which is a double covering group of $Sp_\mathbb{R}(2k)$ [81]

(iv) $trace(A) = \frac{1}{(2\pi\hbar)^k} \int W_A^\hbar(x)dx, \tag{5.9}$

where again, dx stands for the Liouville volume on $\mathbb{R}^{2k}_\omega$, $dx = d^k p d^k q$. Furthermore, Weyl's symbol correspondence in fact also satisfies the stronger property:

$$(v) \quad trace(A^* B) = \frac{1}{(2\pi\hbar)^k} \int \overline{W_A^\hbar}(x) W_B^\hbar(x) dx. \tag{5.10}$$

Clearly, a symbol correspondence satisfying all above properties is a very powerful tool in relating the quantum and classical formalisms of affine mechanical systems, especially because property (v) allows us to compute quantum measurable quantities entirely within the classical formalism of functions on affine symplectic space $\mathbb{R}^{2k}_\omega$ in a way that is equivariant under the full group of symmetries of this space, in accordance with (iii). Furthermore, we can "import" the product of operators to a noncommutative associative product $\star$ of functions on $\mathbb{R}^{2k}_\omega$, by

$$W_{AB}^\hbar = W_A^\hbar \star W_B^\hbar, \tag{5.11}$$

which is called the Weyl product, or the Moyal product, whose commutator is precisely the bracket originally introduced by Moyal [50].

With such a powerful tool in hands, one can thus proceed to study quantum dynamics in the asymptotic limit of high quantum numbers, where classical Poisson dynamics should prevail, entirely within the mathematical framework of functions on $\mathbb{R}^{2k}_\omega$. That is, in this way one first proceeds with a "dequantization" of the quantum mechanical formalism and then studies its semiclassical limit. For affine mechanical systems, this semiclassical limit is often studied by treating $\hbar$ formally as a variable and looking at the asymptotic expressions for the symbols, their products and commutators, measurable quantities, etc., as $\hbar \to 0$.

However, when trying to make a more rigorous mathematical sense of Weyl's symbol correspondence obtained via (5.2)–(5.8), it became clear that one has to be very careful as to which classes of operators on $\mathcal{H}$ and functions on $\mathbb{R}^{2k}_\omega$ should be considered. In fact, although Weyl's correspondence has been presented as a map from operators to functions, in practice equations like (5.8) together with (5.6) have often been used in the opposite direction, that is, of defining new classes of operators, as pseudo-differential or Fourier-integral operators (see Hörmander [27, 39, 40], or in the proper Weyl context see [36] and also [79] for asymptotics). Therefore, this inverse direction of using a symbol correspondence, often also referred to as "quantization", has actually become more familiar to many people.

Moreover, it was soon realized that the symbol correspondence rule obtained via (5.2)–(5.8) is not unique (and it was not the one originally used by Hörmander). In fact, note that choosing $T_0^\hbar(\xi, 1)$ to represent the equivalence class $[T_0^\hbar(\xi)]$ seems to be arbitrary, but more importantly, note also that $\mathbb{R}^{2k}_\omega$ can be more generally "almost identified" with a large family of Heisenberg groups H_α^{2k}, which are defined by modifying the product (5.2) to the more general one given by

$$(\xi_1, \exp(i\theta_1)) \cdot (\xi_2, \exp(i\theta_2)) = (\xi_1 + \xi_2, \exp(i(\theta_1 + \theta_2 + ([\omega + \alpha](\xi_1, \xi_2)/2\hbar)))),$$

where α is a symmetric bilinear form on $\mathbb{R}^{2k}$ and different choices of α are related to different choices of "orderings" for products of operators $\hat{p}_j$ and $\hat{q}_j$. In this respect, Weyl's ordering is the symmetric one, $p_j q_j \leftrightarrow (\hat{p}_j \hat{q}_j + \hat{q}_j \hat{p}_j)/2$, but other popularly used orderings are $p_j q_j \leftrightarrow \hat{p}_j \hat{q}_j$ and $p_j q_j \leftrightarrow \hat{q}_j \hat{p}_j$ (normal ordering). Similarly, one can define different symbol correspondences using complex coordinates $z_j = q_j + i p_j$ and $\bar{z}_j = q_j - i p_j$ on $\mathbb{R}^{2k}_\omega \simeq \mathbb{C}^k$, by considering different orderings for products of $a_j = \hat{q}_j + i\hat{p}_j$ and $a_j^\dagger = \hat{q}_j - i\hat{p}_j$, or by using coherent states, etc.

Furthermore, for affine mechanical systems it is not obvious for which group one should impose equivariance for all possible symbol correspondences. If $W^{\hbar'}$ is another symbol correspondence satisfying (i)–(iv) but not (v), then from (5.11) we see that it also satisfies the weaker property

$$(\mathrm{v}') \quad trace(A^*B) = \frac{1}{(2\pi\hbar)^k} \int \overline{W_A^{\hbar'}} \star W_B^{\hbar'}(x)dx \,, \tag{5.12}$$

so that this property (v') can be used, instead of (v), to compute quantum expectation values within the classical formalism of functions on affine symplectic space. Then, one can consider relaxing (iii) to an equivariance under some subgroup of $aSp_{\mathbb{R}}(2k)$ containing $E_{\mathbb{R}}(k)$ and all affine translations, like $E_{\mathbb{R}}(k \oplus k)$ or $E_{\mathbb{C}}(k)$, as long as (v') is still invariant under the whole affine symplectic group $aSp_{\mathbb{R}}(2k)$ (symbol correspondences define via $\alpha \neq 0$, like Hörmander's normal ordering, are usually not equivariant under the full group $aSp_{\mathbb{R}}(2k)$ in the strong sense of (iii)).

In other words, there are many other symbol correspondences in affine mechanical systems satisfying all or most of properties (i)–(v) above and it would be desirable to classify all such correspondences, study their semiclassical asymptotic limit, see how these correspondences agree or disagree in this limit, etc.

For affine mechanical systems, such a complete and systematic study is not known to us. However, in contrast to quantum affine mechanical systems, for quantum spin-j systems the Hilbert spaces are finite dimensional, allowing for an independent mathematical formulation of such systems, as was done in Chap. 3, and there is never any ambiguity about which classes of operators to consider. Moreover, for spin systems there is no doubt about the natural group of symmetries to impose equivariance: the 3-sphere $SU(2)$, acting effectively via $SO(3)$.

Therefore, going back in time over two millennia, so to speak, and replacing straight lines and k-planes by circles and spheres as the fundamental geometrical objects, we can provide the complete classification and begin a systematic study of symbol correspondences for spin systems, as is presented below in Chaps. 6–8. But it is somewhat curious, perhaps, that while ancient Greek philosophers looked outwards to the far sky in search of circles and spheres, modern physicists found them by looking deeply inwards into matter.

Chapter 6
Symbol Correspondences for a Spin-j System

In Chaps. 3 and 4, quantum spin-j mechanical systems and the classical spin mechanical system, respectively, were defined and studied in fully independent ways. In this chapter, the two formulations are brought together via spin-j symbol correspondences. Inspired by Weyl's correspondence in affine mechanical systems we investigate, for a spin-j system, symbol correspondences that associate operators on Hilbert space to functions on phase space, satisfying certain properties.

Here we define, classify and study such correspondences, presenting explicit constructions. Our cornerstone is the concept of characteristic numbers of a symbol correspondence, cf. Definition 6.2.22, which provides coordinates on the moduli space of spin-j symbol correspondences. As we shall see below, for any j a (quite smaller) subset of characteristic numbers can be distinguished in terms of a stricter requirement for an isometric correspondence. However, a more subtle distinction is obtained in the asymptotic limit $n = 2j \to \infty$, to be explored in Chap. 8. As we shall see there in detail, not all n-sequences of characteristic numbers lead to classical Poisson dynamics in this asymptotic limit.

6.1 General Symbol Correspondences for a Spin-j System

6.1.1 Definition of Spin-j Symbol Correspondences

Following Stratonovich, Varilly and Gracia-Bondia (cf. [67, 73]), in the spirit of Weyl we introduce the following main definition:

Definition 6.1.1. A *symbol correspondence* for a spin-j quantum mechanical system $\mathcal{H}_j \simeq \mathbb{C}^{n+1}$, where $n = 2j$, is a rule which associates to each operator $P \in \mathcal{B}(\mathcal{H}_j)$ a smooth function W_P^j on S^2, called its *symbol*, with the following properties:

© Springer International Publishing Switzerland 2014

P. de M. Rios, E. Straume, *Symbol Correspondences for Spin Systems*,
DOI 10.1007/978-3-319-08198-4_6

(i)   Linearity : The map $P \to W_P^j$ is linear and injective.

(ii)   Equivariance : $W_{Pg}^j = (W_P^j)^g$, for each $g \in SO(3)$.

(iii) Reality : $W_{P*}^j(\mathbf{n}) = \overline{W_P^j(\mathbf{n})}$.   (6.1)

(iv)   Normalization : $\frac{1}{4\pi} \cdot \int\limits_{S^2} W_P^j dS = \frac{1}{n+1} trace(P)$.

By dropping the injectivity requirement in (i), any map

$$W^j : \mathcal{B}(\mathcal{H}_j) \to C_{\mathbb{C}}^{\infty}(S^2),$$

satisfying properties (i)–(iv), except that it may not be injective, shall be called a *pre-symbol map*. In this way, the symbol correspondences characterize a distinguished family of pre-symbol maps.

On the other hand, the injectivity requirement in (i) can be transformed to a bijectivity requirement by reducing the target space of every map

$$W^j : \mathcal{B}(\mathcal{H}_j) \simeq M_{\mathbb{C}}(n+1) \to Poly_{\mathbb{C}}(S^2)_{\leq n} \subset C_{\mathbb{C}}^{\infty}(S^2)   \qquad (6.2)$$

which associates to each operator P its *symbol* W_P^j. We shall assume *the symbols W_P^j are polynomial functions*, as indicated in (6.2), unless otherwise stated.

Remark 6.1.2. In addition to the above four axioms (6.1) one may also impose the axiom

$$(v) \quad \text{Isometry}: \left\langle W_P^j, W_Q^j \right\rangle = \langle P, Q \rangle_j   \qquad (6.3)$$

as a "metric normalization" condition, where the right-hand side of Eq. (6.3) is the normalized Hilbert-Schmidt inner product of two operators, given by

$$\langle P, Q \rangle_j = \frac{1}{n+1} \langle P, Q \rangle = \frac{1}{n+1} trace(P^*Q) ,   \qquad (6.4)$$

$n = 2j$, and the left-hand side of (6.3) is the normalized L^2 inner product of two functions on the sphere, given by (4.4). Thus, condition (iv) is just a special case of (v), namely (iv) can be stated as

$$(iv)' \quad \left\langle 1, W_P^j \right\rangle = \langle I, P \rangle_j .$$

Berezin [10–12] was the first to investigate symbol correspondences for spin-j systems in a more systematic way. His correspondence satisfies axioms (i)–(iv), but not axiom (v). Varilly and Gracia-Bondia [73] were the first to systematically investigate the rules $P \mapsto W_P^j$ satisfying all five axioms, as previously outlined by Stratonovich [67] for the spin-j-system version of the Weyl correspondence.

Definition 6.1.3. A *Stratonovich-Weyl correspondence* is a symbol correspondence that also satisfies the isometry axiom (v).

Remark 6.1.4. **Varilly and Gracia-Bondia's** justification for axiom (v) is the need "to assure that quantum mechanical expectations can be calculated by taking integrals over the sphere". However, this justification is not entirely proper, as these latter integrals can be defined for any symbol correspondence (cf. Eq. (6.25) and Remark 6.2.25, as well as Remark 7.1.4 and Eq. (7.3), below). This situation mimics, in fact, the situation for symbol correspondences in affine mechanical systems: while Weyl's symbol correspondence satisfies property (v) given in (5.10), many other useful symbol correspondences in affine mechanical systems do not.

6.1.2 The Moduli Space of Spin-j Symbol Correspondences

We recall that the action of $SU(2)$ on operators in $M_{\mathbb{C}}(n+1)$, as explained in Chap. 2, is by conjugation via the representation $\varphi_j : SU(2) \to SU(n+1)$,

$$g \in SU(2) : A \to A^g = \varphi_j(g)A\varphi_j(g^{-1}),$$

which factors to an effective action of $SO(3)$, while the action of $SU(2)$ on polynomial functions in $Poly_{\mathbb{C}}(S^2)_{\leq n}$, as explained in Chap. 3, is obtained from the standard action of $SO(3)$ on the two-sphere:

$$g \in SU(2) : F \to F^g , \ F^g(\mathbf{n}) = F(g^{-1}\mathbf{n}),$$

so that in both spaces we have an effective left action of $G = SO(3)$.

Furthermore, by Proposition 4.2.2 these two spaces are isomorphic and have the same splitting into G-invariant subspaces, both in the complex and the real case. Therefore, by the classical Schur's lemma (Lemma 2.1.3) applied to (4.23),

$$Hom^G(M_{\mathbb{C}}(n+1), Poly_{\mathbb{C}}(S^2)_{\leq n}) \simeq \prod_{l=0}^{n} Hom^G(M_{\mathbb{C}}(\varphi_l), Poly(\varphi_l)) \simeq \mathbb{C}^{n+1},$$

(6.5)

and similarly, by restricting to the real case (4.22),

$$Hom^G(\mathcal{H}(n+1), Poly_{\mathbb{R}}(S^2)_{\leq n}) \simeq \prod_{l=0}^{n} Hom^G(\mathcal{H}(\psi_l), Poly(\psi_l)) \simeq \mathbb{R}^{n+1}. \quad (6.6)$$

Now, note that (i) + (ii) in Definition 6.1.1 is the condition that W^j is a G-isomorphism, and (iii) assures that W^j preserves the real structure. Therefore, we have the following:

Corollary 6.1.5. *Each $\mathbb{R}$-linear G-map which takes Hermitian matrices to real polynomials may be identified with a unique $(n+1)$-tuple*

$$(c_0, c_1, \ldots, c_n) \in \mathbb{R}^{n+1}. \tag{6.7}$$

In particular, the tuple corresponds to a G-isomorphism

$$\mathcal{H}(n+1) \simeq Poly_{\mathbb{R}}(S^2)_{\leq n},$$
$$\cap \qquad\qquad \cap \tag{6.8}$$
$$M_{\mathbb{C}}(n+1) \simeq Poly_{\mathbb{C}}(S^2)_{\leq n},$$

if and only if each $c_l \neq 0$, $l = 0, 1, \ldots, n$.

Remark 6.1.6. The above correspondence is not canonical since there is no natural choice of isomorphism by which one may identify matrices with functions on S^2 equivariantly. This multitude of choices is a central topic in the next section, where the numbers c_l will be defined in a precise way.

But note also that the symbol W_I^j of the identity operator is a constant function, say equal to c_0, and then condition (iv) clearly implies

$$c_0 = 1, \tag{6.9}$$

namely, the symbol map respects the unit elements of the two rings in (6.2). In this way, we can identify each W^j by its real n-tuple representation:

$$W^j \leftrightarrow (c_1, \ldots, c_n) \in (\mathbb{R}^*)^n, \text{ where } \mathbb{R}^* = \mathbb{R} - \{0\}, \tag{6.10}$$

(cf. Remark 6.1.6). For an explicit definition of the numbers c_l we refer to Theorem 6.2.7 and Definition 6.2.22 below. To summarize, we have obtained the following:

Theorem 6.1.7. *The moduli space of all spin-j symbol correspondences satisfying conditions (i)–(iv) in Definition 6.1.1 is $(\mathbb{R}^*)^n$.*

Note that nothing has yet been said with respect to the isometric spin-j symbol correspondences, i.e. the Stratonovich-Weyl symbol correspondences, which also satisfy condition (v).

In our general setting to be developed below, we fix a representation (6.10), either by coupled basis decomposition or more systematically in terms of operator kernels, and relate the metric properties of W^j to the numbers c_l (further below, we also present a direct way to define the symbol map introduced by Berezin).

Then, it will be seen below that condition (v) determines each c_l up to sign.

6.2   Explicit Constructions of Spin-j Symbol Correspondences

We consider the general category of spin-j symbol correspondences.

Remark 6.2.1. The objects of this category are the symbol correspondences, while the morphisms are induced from the transition operators to be defined later, cf. Definition 7.1.18. Because all morphisms in this (small) category are invertible, this is actually a *groupoid*, cf. Definition 7.1.19.

As a basis point on S^2 we choose the north pole $\mathbf{n}_0 = (0, 0, 1)$, the positive z direction (cf. Definition 3.1.4 and Remark 4.2.1), assuming that its isotropy group is the circle group $U(1) \subset SU(2)$ in (2.37) whose fixed point set in $M_{\mathbb{C}}(n + 1)$ consists of the diagonal matrices, namely the $m = 0$ eigenspace of $J_3 = J_{\mathbf{n}_0}$ (cf. Eq. (3.34) and the discussion following Definition 3.3.1).

From now on, we shall often write W for W^j, for simplicity, whenever suppressing the spin number $j = n/2$ is not a cause for confusion.

6.2.1 Symbol Correspondences via Standard Basis

Following on the same reasoning that led to Corollary 6.1.5 above, a natural and simple way to establish a symbol correspondence between operators and polynomial functions is obtained by relating appropriately chosen basis for each space.

Thus, remind that for a given $n = 2j$, the operator space $\mathcal{B}(\mathcal{H}_j)$ has the orthogonal decomposition

$$M_{\mathbb{C}}(n + 1) = \sum_{l=0}^{n} M_{\mathbb{C}}(\varphi_l)$$

where each summand $M_{\mathbb{C}}(\varphi_l)$ has its standard basis $\mathbf{e}^j(l, m), -l \leq m \leq l$, in accordance with Proposition 3.3.5 and Theorem 3.3.6.

On the other hand, the space of polynomial functions on S^2 of proper degree $\leq n$ has the orthogonal decomposition

$$Poly_{\mathbb{C}}(S^2)_{\leq n} = \sum_{l=0}^{n} Poly(\varphi_l)$$

where each summand $Poly(\varphi_l)$ of polynomials of proper degree l has its standard basis of spherical harmonics $Y_{l,m}, -l \leq m \leq l$.

Consequently, for a given value of n and signs $\varepsilon_l^n = \pm 1, l = 1, 2, \ldots, n$, we can set up the 1-1 correspondence

$$\mu_0 \mathbf{e}^j(l, m) \longleftrightarrow \varepsilon_l^n Y_{l,m} \; ; \; -l \leq m \leq l \leq n, \; \mu_0 = \sqrt{n + 1}. \qquad (6.11)$$

Such a correspondence is obviously isometric, therefore, if it extends linearly to a symbol correspondence in the sense of Definition 6.1.1, then clearly all the 2^n Stratonovich-Weyl symbol correspondences are obtained in this way. In fact, by relaxing on the isometric requirement and allowing for scaling freedom, we have the more general result, whose formal proof is deferred to the next section:

Proposition 6.2.2. *Any symbol correspondence W^j satisfying Definition 6.1.1 is uniquely determined by non-zero real numbers c_l^n, $l = 1, 2, \ldots, n$, which yield the explicit 1-1 correspondence*

$$W^j = W_{\vec{c}}^j : \mu_0 e^j (l, m) \mapsto c_l^n Y_{l,m} \; ; \; -l \le m \le l \le n, \; c_l^n \in \mathbb{R}^*, \tag{6.12}$$

where $\vec{c}$ is a shorthand notation for the n-string $(c_1^n, c_2^n, \ldots, c_n^n)$. Furthermore, W^j is a Stratonovich-Weyl symbol correspondence if and only if $c_l^n = \varepsilon_l^n = \pm 1$, $l = 1, 2, \ldots, n$.

Remark 6.2.3. We can now understand better the moduli space of symbol correspondences W. Starting with the isometric ones, recall from Remark 4.2.4 that each respective standard basis of $M_{\mathbb{C}}(\varphi_l)$ and $Poly(\varphi_l)$ is uniquely defined up to an arbitrary overall phase. Therefore, an isometric correspondence between $M_{\mathbb{C}}(\varphi_l)$ and $Poly(\varphi_l)$ is uniquely defined modulo a relative phase $z_l \in S^1 \subset \mathbb{C}$. However, the reality requirement (iii) in Definition 6.1.1 fixes this phase to be real, thus $z_l = \varepsilon_l = \pm 1$. For the non-isometric correspondences, we have the further freedom of a relative scaling $\rho_l \in \mathbb{R}^+$ and, therefore, a correspondence between $M_{\mathbb{C}}(\varphi_l)$ and $Poly(\varphi_l)$ is uniquely defined modulo a number $c_l = \rho_l \varepsilon_l \in \mathbb{R}^*$.

6.2.2 Symbol Correspondences via Operator Kernel

In order to study general symbol correspondences more systematically, observe that a diagonal matrix K gives rise to a function $K(\mathbf{n})$ on S^2 such that $K(\mathbf{n}_0) = K$, and $K(\mathbf{n}) = K^g$ for $\mathbf{n} = g\mathbf{n}_0$.

Proposition 6.2.4. *For each symbol correspondence $W = W^j$ there is a unique operator $K \in M_{\mathbb{C}}(n + 1)$ such that*

$$W_P(g\mathbf{n}_0) = trace(PK^g) \tag{6.13}$$

or equivalently,

$$W_P(\mathbf{n}) = trace(PK(\mathbf{n})) = \langle P^*, K(\mathbf{n}) \rangle .$$

Moreover, K is a diagonal matrix with real entries and trace 1.

Proof. The linear functional

$$\hat{W} : M_{\mathbb{C}}(n + 1) \to \mathbb{C}, \; P \mapsto W_P(\mathbf{n}_0)$$

is represented by some K such that $\hat{W}(P) = trace(PK)$, since the pairing $\langle K^*, P \rangle = trace(PK)$ is a Hermitian inner product. Therefore, (6.13) holds for $g = 1$ and hence also in general by equivariance

$$W_P(g^{-1}\mathbf{n}_0) = (W_P)^g(\mathbf{n}_0) = W_{P^g}(\mathbf{n}_0) = trace(P^g K) = trace(PK^{g^{-1}}).$$

On the other hand, for $g \in U(1)$ we have $g\mathbf{n}_0 = \mathbf{n}_0$ and then $trace(PK^g) = trace(PK)$ holds for each P, consequently $K^g = K$. That is, K is fixed by $U(1)$ and hence $K = diag(\lambda_1, \ldots, \lambda_{n+1})$. By choosing P to be the one-element matrix $\mathcal{E}_{kk}$ it follows that $\lambda_k = W_P(\mathbf{n}_0)$ is a real number, due to the reality condition (iii). Finally, by the normalization condition (iv) $W_I = 1$ and hence $trace(K) = 1$. $\square$

Definition 6.2.5. An operator kernel $K \in M_{\mathbb{C}}(n + 1)$ is a diagonal matrix with the property that the symbol map W defined by (6.13) is a symbol correspondence.

It follows from Proposition 6.2.4 that K has an orthogonal decomposition

$$K = \frac{1}{n + 1} I + K_1 + K_2 + \cdots + K_n \tag{6.14}$$

where for $l \geq 1$ each K_l is a zero trace real diagonal matrix belonging to the zero weight (or $m = 0$) subspace of the irreducible summand $M_{\mathbb{C}}(\varphi_l)$ of the full matrix space (3.56), namely $K_l = k_l e(l, 0)$ for some non-zero real number k_l.

Conversely, each matrix K of this kind defines a symbol map $P \to W_P$ by (6.13) whose kernel (as a linear map) is a G-invariant subspace, namely the sum of those $M_{\mathbb{C}}(\varphi_l)$ for which $K_l = 0$. Therefore, by axiom (i) in (6.1) each K_l must be non-zero for a symbol correspondence.

Lemma 6.2.6. *Let K be a real diagonal matrix with $trace(K) = 1$, and define a symbol map W by the formula (6.13). Then W satisfies the normalization condition, namely*

$$\frac{1}{4\pi} \int_{S^2} W_P dS = \frac{1}{n + 1} trace(P). \tag{6.15}$$

Proof. Let (6.14) be the orthogonal decomposition of K, where $K_l \in M_{\mathbb{C}}(\varphi_l)$ may possibly be zero, and let dg denote the normalized measure on $SU(2)$. Then

$$\frac{1}{4\pi} \int_{S^2} W_P dS = \int_{SU(2)} W_P(g\mathbf{n}_0) dg = \int_{SU(2)} trace(PK^g) dg =$$

$$= trace\left(P \int_{SU(2)} K^g dg\right) = trace\left[P \left(\frac{1}{n+1} I + \sum_{l=1}^{n} \left(\int_{SU(2)} \tilde{\varphi}_l(g) dg\right) K_l\right)\right],$$

where the operator $\tilde{\varphi}_l(g) \in GL(M_{\mathbb{C}}(\varphi_l))$ is the action of g on the vector space $M_{\mathbb{C}}(\varphi_l)$, in particular

$$\tilde{\varphi}_l(g)K_l = K_l^g = \varphi_l(g)K_l\varphi_l(g)^{-1.}$$

Since the representation $g \to \tilde{\varphi}_l(g)$ is irreducible, it follows by standard representation theory that

$$\int_{SU(2)} \tilde{\varphi}_l(g)dg = 0,$$

and this proves the identity (6.15).   □

Putting together the above results yields the following classification of all possible symbol correspondences:

Theorem 6.2.7. *The construction of symbol maps $W = W^j$ in terms of operator kernels $K = K^j$ by the formula (6.13) establishes a bijection between symbol correspondences and real diagonal matrices in $M_{\mathbb{C}}(n+1)$ of type*

$$K^j = \frac{1}{n+1}I + \sum_{l=1}^{n} c_l \sqrt{\frac{2l+1}{n+1}} \mathbf{e}^j(l,0), \quad c_l \neq 0 \; real, \tag{6.16}$$

where $\mathbf{e}^j(l,0) \in M_{\mathbb{C}}(\varphi_l)$ is the traceless diagonal matrix of unit norm given by

$$\mathbf{e}^j(l,0) = \frac{(-1)^l}{l!}\sqrt{2l+1}\sqrt{\frac{(n-l)!}{(n+l+1)!}}\sum_{k=0}^{l}(-1)^k \binom{l}{k} J_-^{l-k} J_+^l J_-^k \tag{6.17}$$

(cf. Theorem 3.3.6). In particular, the symbol correspondence W^j defined by formula (6.13) is determined by the n-tuple $(c_1, c_2, \ldots, c_n) \in (\mathbb{R}^*)^n$.

Definition 6.2.8. The non-zero real numbers $c_1, c_2, \ldots, c_n$ will be referred to as the *characteristic numbers* of the operator kernel K.

In the general context of Theorem 6.2.7, we now distinguish the symbol correspondences of the kind originally defined by Berezin, as follows.

Definition 6.2.9. A *Berezin symbol correspondence* is a symbol correspondence whose operator kernel K is a projection operator Π (cf. e.g. [57]).

Since the trace of a projection operator is its rank, it follows that Π must be a one-element matrix $\Pi_k = \mathcal{E}_{kk}$ for some $1 \leq k \leq n + 1 = 2j + 1$. Namely, the expansion (6.16) of Π_k in the coupled standard basis reads

$$\Pi_k = (-1)^{k+1}|jmj(-m)\rangle = \frac{1}{n+1}I + (-1)^{k+1}\sum_{l=1}^{n} C_{m,-m,0}^{j,j,l}\mathbf{e}(l,0), \tag{6.18}$$

$$\text{where} \quad m = j - k + 1,$$

(cf. (3.32) and (3.40)) and hence each Clebsch-Gordan coefficient in the above sum must be non-zero in the Berezin case. We state this result as follows:

Proposition 6.2.10. *For a spin-j quantum system $\mathcal{H}_j = \mathbb{C}^{n+1}$, the Berezin symbol correspondences are characterized by having as operator kernel a projection $\Pi_k = \mathcal{E}_{kk}$, $1 \leq k \leq n+1$, for those k such that the following Clebsch-Gordan coefficients are non-zero:*

$$C^{j,\,j,l}_{m,-m,0} \neq 0, \quad m = j - k + 1, 1 \leq l \leq n = 2j. \tag{6.19}$$

Equivalently, the kth entry of each diagonal matrix $\mathbf{e}(l,0)$, $l = 1,2,\ldots,n$, must be non-zero.

Remark 6.2.11. The traceless matrix $\mathbf{e}(l,0)$ has the following symmetry

$$\mathbf{e}(l,0) = diag(d_1, d_2, \ldots, d_{n+1}), \quad d_i = \pm d_{n+2-i}.$$

When $n = 2j$ is odd, namely half-integral spin j, we conjecture that the Berezin condition (6.19) holds for each k. We have checked this for all $n < 20$ say, since for these cases the above matrices $\mathbf{e}(l,0)$ can be calculated easily using computer algebra. However, for integral values of j the Berezin condition fails for the projection operator Π_k when $k = n/2 + 1$. Moreover, $d_2 = 0$ holds for $n = l(l+1) = 6, 12, 20, \ldots$, so in these dimensions the Berezin condition also fails for Π_2 and Π_n. We will come back to this discussion in Sect. 6.2.4.

On the other hand, $d_1 \neq 0$ always holds, and consequently the Berezin condition (6.19) is satisfied for $k = 1$ and $k = n+1$. Therefore, for all $n = 2j \in \mathbb{N}$, the operators Π_1 and Π_{n+1} always yield Berezin symbol correspondences.

Definition 6.2.12. The symbol obtained via the projection operator Π_1 will be called the *standard Berezin symbol*.

Remark 6.2.13. The standard Berezin symbol correspondence generalizes to spin systems the method of correspondence via "coherent states" originally introduced for affine quantum mechanics [33, 70]. In fact, this method can be applied in more general settings (cf. [53]) and the original papers by Berezin were already cast in the more general context of complex symmetric spaces [10–12]. Actually, the originally defined coherent state in affine quantum mechanics is one of many, so it can be called the standard one. Similarly, we shall call Berezin's the standard coherent state of a quantum spin-j system when we generalize it, in Sect. 6.2.4.

General Metric Relation

Now, the following proposition sets a general "metric" relation, similar to axiom (v) in Remark 6.1.2 which is valid for all symbol correspondences.

Let K in (6.14) and (6.16) be given, write $\mathbf{e}(0,0) = \frac{1}{\sqrt{n+1}} I$, let $P \in M_{\mathbb{C}}(n+1)$ and consider the orthogonal decompositions

$$P = \sum_{l=0}^{n} P_l = \sum_{l=0}^{n} \sum_{m=-l}^{l} a_{lm} \mathbf{e}(l,m), \quad K = \sum_{l=0}^{n} K_l = \sum_{l=0}^{n} \gamma_l \mathbf{e}(l,0) . \qquad (6.20)$$

Proposition 6.2.14. *Any symbol correspondence W satisfies the metric identity*

$$\langle W_P, W_Q \rangle = \sum_{l=0}^{n} \frac{\gamma_l^2}{2l+1} \langle P_l, Q_l \rangle = \sum_{l=0}^{n} \frac{(c_l)^2}{n+1} \langle P_l, Q_l \rangle \qquad (6.21)$$

where the γ_l and c_l are related by

$$\gamma_l = c_l \sqrt{\frac{2l+1}{n+1}}, \; \text{cf. (6.16) and (6.20), Theorem 6.2.7.}$$

Proof. From (6.13) and (6.20), we have that

$$W_P(g\mathbf{n}_0) = \sum_{l=0}^{n} trace(P_l K^g) = \sum_{l=0}^{n} \sum_{l'=0}^{n} \sum_{m=-l}^{l} a_{lm} trace(\mathbf{e}(l,m) K_{l'}^g)$$

$$= \sum_{l=0}^{n} \sum_{l'=0}^{n} \sum_{m=-l}^{l} a_{lm} \gamma_{l'} trace(\mathbf{e}(l,m) \mathbf{e}(l',0)^g) = \sum_{l=0}^{n} \sum_{m=-l}^{l} (-1)^m a_{lm} \gamma_l D_{-m,0}^l(g)$$

where the inner product

$$D_{-m,0}^l(g) = \langle \mathbf{e}(l,-m), \mathbf{e}(l,0)^g \rangle = trace((-1)^m \mathbf{e}(l,m) \mathbf{e}(l,0)^g)$$

is a Wigner D-function, namely a matrix element of the unitary matrix $D^l(g)$ representing the action of g on the irreducible operator subspace $M_{\mathbb{C}}(\varphi_l) \simeq \mathbb{C}^{2l+1}$. Consequently, expanding $Q \in M_{\mathbb{C}}(n+1)$ similarly to (6.20) we obtain

$$\frac{1}{4\pi} \int_{S^2} W_{P^*}(\mathbf{n}) W_Q(\mathbf{n}) dS = \int_G W_{P^*}(g\mathbf{n}_0) W_Q(g\mathbf{n}_0) dg$$

$$= \sum_{l=0}^{n} \sum_{l'=0}^{n} \sum_{m=-l}^{l} \sum_{m'=-l'}^{l'} \bar{a}_{lm} b_{l'm'} \gamma_l \gamma_{l'} \int_G \overline{D_{-m,0}^l(g)} D_{-m',0}^{l'}(g) dg$$

$$= \sum_{l=0}^{n} \sum_{m=-l}^{l} \bar{a}_{lm} b_{lm} \gamma_l^2 \int_G \left| D_{-m,0}^l(g) \right|^2 dg$$

$$= \sum_{l=0}^{n} \sum_{m=-l}^{l} \bar{a}_{lm} b_{lm} \frac{\gamma_l^2}{2l+1} = \sum_{l=0}^{n} trace(P_l^* Q_l) \frac{\gamma_l^2}{2l+1}$$

where we have used the well-known Frobenius-Schur orthogonality relations for the matrix elements $D^l_{m,m'}(g)$ of irreducible unitary representations. $\qquad\square$

From (6.21) we also deduce the formula

$$(c_l)^2 = (n+1)\frac{\|W_P\|^2}{\|P\|^2}, \text{ for any non-zero } P \in M_{\mathbb{C}}(\varphi_l). \tag{6.22}$$

Corollary 6.2.15. *For each j, W^j is a Stratonovich-Weyl symbol correspondence if and only if the characteristic numbers are*

$$c_l = \varepsilon_l = \pm 1, \, l = 1, \ldots, n = 2j \tag{6.23}$$

and consequently there are precisely 2^n different symbol maps W^j of this type, in agreement with Theorem 1 of [73].

Remark 6.2.16. Formula (6.22) also gives $(c_0)^2 = 1$, but the value $c_0 = -1$ in (6.23) is already excluded by the normalization axiom (iv), cf. (6.1) and (6.9).

Summary 6.2.17. *It follows from (6.21) that each symbol correspondence becomes an isometry by appropriately scaling the (Hilbert-Schmidt) inner product on each of the irreducible matrix subspaces $M_{\mathbb{C}}(\varphi_l)$. The moduli space of all symbol correspondences is $(\mathbb{R}^*)^n$, having 2^n connected components, and each symbol correspondence can be continuously deformed to a unique Stratonovich-Weyl symbol correspondence in the moduli space $(\mathbb{Z}_2)^n$.*

Covariant-Contravariant Duality

Given an operator kernel K in the sense of Definition 6.2.5 and Theorem 6.2.7, it is also possible to define a symbol correspondence $\tilde{W}$ via the integral equation

$$P = \frac{n+1}{4\pi}\int_{S^2} \tilde{W}_P(g\mathbf{n}_0)K^g dS = \frac{n+1}{4\pi}\int_{S^2} \tilde{W}_P(\mathbf{n})K(\mathbf{n})dS \tag{6.24}$$

where $\mathbf{n} = g\mathbf{n}_0 \in S^2 = SO(3)/SO(2)$ and $K^g = K(\mathbf{n})$, cf. also (6.13).

Definition 6.2.18. The symbol map

$$\tilde{W} = \tilde{W}^{j,K} : \mathcal{B}(\mathcal{H}_j) \simeq M_{\mathbb{C}}(n+1) \to Poly_{\mathbb{C}}(S^2)_{\leq n} \subset C^{\infty}_{\mathbb{C}}(S^2)$$

defined implicitly by Eq. (6.24) is called the *contravariant* symbol correspondence given by the operator kernel K. On the other hand, the symbol map

$$W = W^{j,K} : \mathcal{B}(\mathcal{H}_j) \simeq M_{\mathbb{C}}(n+1) \to Poly_{\mathbb{C}}(S^2)_{\leq n} \subset C^{\infty}_{\mathbb{C}}(S^2)$$

defined explicitly by Eq. (6.13) is called the *covariant* symbol correspondence given by the operator kernel K.

Remark 6.2.19. This terminology of covariant and contravariant symbol correspondences was introduced by Berezin [12]. See also Remark 6.2.25, below.

Proposition 6.2.20. *Let K be the operator kernel for a Stratonovich-Weyl covariant symbol correspondence $W^{j,K}$, that is, $c_l = \pm 1$, $1 \leq l \leq n$. Then, $\tilde{W}^{j,K} \equiv W^{j,K}$, that is, for any operator $P \in B(\mathcal{H}_j)$, $\tilde{W}_P \equiv W_P \in C_{\mathbb{C}}^{\infty}(S^2)$. In other words, for a Stratonovich-Weyl symbol correspondence, the covariant and the contravariant symbol correspondences (defined by K) coincide.*

Proof. From (6.24) and (6.13), we have that

$$trace(PQ) = \frac{n+1}{4\pi} \int_{S^2} \tilde{W}_P(\mathbf{n}) W_Q(\mathbf{n}) dS , \qquad (6.25)$$

but, since $W^{j,K}$ is an isometry (cf. (6.4)),

$$trace(PQ) = \frac{n+1}{4\pi} \int_{S^2} W_P(\mathbf{n}) W_Q(\mathbf{n}) dS .$$

Since both equations are valid $\forall P, Q \in B(\mathcal{H}_j)$, it follows that $\tilde{W}_P \equiv W_P$.   □

On the other hand, by analogous reasoning from Eqs. (6.24) and (6.13) and the metric identity (6.21), there is the following more general result:

Theorem 6.2.21. *Let K be determined by the characteristic numbers $c_1, c_2, \ldots, c_n$, as explained by Theorem 6.2.7, and let*

$$W^{j,K} : B(\mathcal{H}_j) \simeq M_{\mathbb{C}}(n+1) \to Poly_{\mathbb{C}}(S^2)_{\leq n} \subset C_{\mathbb{C}}^{\infty}(S^2)$$

be the covariant symbol correspondence defined explicitly by Eq. (6.13). Then,

$$W^{j,K} \equiv \tilde{W}^{j,\tilde{K}} : B(\mathcal{H}_j) \to C_{\mathbb{C}}^{\infty}(S^2) , \qquad (6.26)$$

where

$$\tilde{W}^{j,\tilde{K}} : B(\mathcal{H}_j) \simeq M_{\mathbb{C}}(n+1) \to Poly_{\mathbb{C}}(S^2)_{\leq n} \subset C_{\mathbb{C}}^{\infty}(S^2)$$

is the contravariant symbol correspondence defined implicitly by equation

$$P = \frac{n+1}{4\pi} \int_{S^2} \tilde{W}_P(\mathbf{n}) \tilde{K}^g dS , \qquad (6.27)$$

with the operator kernel $\tilde{K}$ determined by the characteristic numbers $\tilde{c}_1, \tilde{c}_2, \ldots, \tilde{c}_n$, where

$$\tilde{c}_l = \frac{1}{c_l} . \qquad (6.28)$$

In other words, (6.27) and (6.28) hold iff $\tilde{W}_P^{\tilde{K}}(\mathbf{n}) = trace(PK^g) = W_P^K(\mathbf{n})$.

As a direct consequence of the above theorem, the relation between the covariant symbol correspondence W given by an operator kernel K and the associated contravariant symbol correspondence $\tilde{W}$ given by the same operator kernel K can be expressed as follows. For any $P = \sum_{l=0}^{n} P_l \in \mathcal{B}(\mathcal{H}_j)$, we have

$$\tilde{W}_P = \sum_{l=0}^{n} \tilde{W}_{P_l} = \sum_{l=0}^{n} \frac{1}{(c_l)^2} W_{P_l}, \tag{6.29}$$

$$W_P = \sum_{l=0}^{n} W_{P_l} = \sum_{l=0}^{n} (c_l)^2 \tilde{W}_{P_l}. \tag{6.30}$$

Definition 6.2.22. The non-zero real numbers $c_1, c_2, \ldots, c_n$, which are the characteristic numbers of the operator kernel K defining the covariant symbol correspondence $W : \mathcal{B}(\mathcal{H}_j) \to C_{\mathbb{C}}^{\infty}(S^2)$ explicitly by Eq. (6.13), will also be referred to as *the characteristic numbers of the symbol correspondence* W.

Definition 6.2.23. A symbol correspondence is *characteristic-positive* if all $c_l > 0$.

Definition 6.2.24. The unique characteristic-positive Stratonovich-Weyl symbol correspondence, for which all characteristic numbers are 1, i.e. $c_l = 1$, $l = 1, \ldots, n$, is called the *standard Stratonovich-Weyl symbol correspondence* and is denoted by

$$W_1^j : \mathcal{B}(\mathcal{H}_j) \simeq M_{\mathbb{C}}(n+1) \to Poly_{\mathbb{C}}(S^2)_{\leq n} \subset C_{\mathbb{C}}^{\infty}(S^2).$$

Remark 6.2.25. There is a *duality* $W \longleftrightarrow \tilde{W}$ between symbol correspondences, namely for a given K the covariant symbol correspondence W^K and the contravariant symbol correspondence $\tilde{W}^K$ are dual to each other. According to Theorem 6.2.21, the passage to the dual symbol correspondence is described by inverting the characteristic numbers, that is, by the replacement $c_i \longrightarrow c_i^{-1}$. Thus, if K (resp. $\tilde{K}$) has characteristic numbers $\{c_i\}$ (resp. $\{c_i^{-1}\}$), then $\tilde{W}^K = W^{\tilde{K}}$ and, as observed in Theorem 6.2.21, $\tilde{W}^{\tilde{K}}$ coincides with W^K. The Stratanovich-Weyl symbol correspondences are precisely the self-dual correspondences for a spin-j system.

We now turn to the formal proof of Proposition 6.2.2 of Sect. 6.2.1.

Proof of Proposition 6.2.2. To see why the choice $c_l^n \equiv c_l$ in (6.12) yields an operator kernel K given by the formula (6.16), we apply formula (6.13) with $P = \mathbf{e}(l, 0)$ and $g = e$, and use the fact that $Y_{l,0}$ takes the value $\sqrt{2l+1}$ at the north pole $\mathbf{n}_0 = (0, 0, 1)$. The second part of the proposition is immediate (cf. Corollary 6.2.15). □

Remark 6.2.26. The numbers $c_l \equiv c_l^n$ of Proposition 6.2.2 are precisely the characteristic numbers of W in the sense of Definition 6.2.22. The notation c_l^n instead of c_l is to emphasize their dependence on $n = 2j$, whenever this dependence is an important issue. We also denote by $\vec{c}$ the n-string $(c_1, c_2, \ldots, c_n) \equiv (c_1^n, c_2^n, \ldots, c_n^n)$, as in (6.12), and by $\frac{1}{\vec{c}}$ the n-string $(\frac{1}{c_1}, \frac{1}{c_2}, \ldots, \frac{1}{c_n})$. The notation $W_{\vec{c}}$ for W^K will be heavily used in the sequel. In particular, the dual of $W_{\vec{c}}$ is $\tilde{W}_{\vec{c}} = W_{\frac{1}{\vec{c}}}$.

6.2.3 Symbol Correspondences via Hermitian Metric

Let $n = 2j$, as usual. We shall construct a symbol map

$$B : \mathcal{B}(\mathcal{H}_j) \simeq M_{\mathbb{C}}(n+1) \to Poly_{\mathbb{C}}(S^2)_{\leq n} \subset C_{\mathbb{C}}^{\infty}(S^2)$$

with the appropriate properties, using the Hermitian metric on the underlying Hilbert space $\mathcal{H}_j \simeq \mathbb{C}^{n+1}$, which we may take to be the space of binary n-forms.

First of all, consider the following explicit construction of a map:

$$\Phi_j : \mathbb{C}^2 \to \mathbb{C}^{n+1}, \quad \mathbb{C}^2 \supset S^3(1) \to S^{2n+1}(1) \subset \mathbb{C}^{n+1}, \tag{6.31}$$

$$\mathbf{z} = (z_1, z_2) \mapsto \Phi_j(\mathbf{z}) = \tilde{Z} = (z_1^n, \sqrt{\binom{n}{1}} z_1^{n-1} z_2, \ldots, \sqrt{\binom{n}{k}} z_1^{n-k} z_2^k, \ldots, z_2^n)$$

where the components of $\tilde{Z}$ can be regarded as normalized binary n-forms, so that, for $S^3(1)$ being the unit sphere in $\mathbb{C}^2$, $S^{2n+1}(1)$ is the unit sphere in $\mathbb{C}^{n+1}$ (cf. the discussion at the end of Sect. 2.2, in particular Eq. (2.44)). As a consequence, as $SU(2)$ acts on $\mathbb{C}^2$ by the standard representation $\varphi_{1/2}$, the induced action on n-forms is the irreducible unitary representation

$$\varphi_j : SU(2) \to SU(n+1)$$

and the map Φ_j is φ_j-equivariant. Next, consider the Hopf map

$$\pi : S^3(1) \to S^2 \simeq \mathbb{C}P^1, \pi(\mathbf{z}) = [z_1, z_2] = \mathbf{n}, \tag{6.32}$$

also described in (2.31), which is equivariant when $SU(2)$ acts on S^2 by rotations via the induced homomorphism $\psi : SU(2) \to SO(3)$, cf. Sect. 2.2.

And finally, let $h : \mathbb{C}^{n+1} \times \mathbb{C}^{n+1} \to \mathbb{C}$ be the usual Hermitian inner product which is conjugate linear in the first variable. Then we have the following:

Theorem 6.2.27. *The map B that associates to each operator $P \in M_{\mathbb{C}}(n+1)$ the function B_P on S^2 defined by*

$$B_P(\mathbf{n}) = h(\tilde{Z}, P\tilde{Z}), \tag{6.33}$$

is a symbol correspondence, according to Definition 6.1.1, whose operator kernel, according to Definition 6.2.5, is the projection operator Π_1.

The proof of Theorem 6.2.27 follows from the set of lemmas below:

Lemma 6.2.28. *The function B_P is well defined and $SU(2)$-equivariant.*

Proof. First, note that B_P is well defined because

$$\Phi(e^{i\theta}\mathbf{z}) = e^{in\theta}\tilde{Z}, \quad h(e^{in\theta}\tilde{Z}, Pe^{in\theta}\tilde{Z}) = h(e^{in\theta}\tilde{Z}, e^{in\theta}P\tilde{Z}) = h(\tilde{Z}, P\tilde{Z}).$$

By the identity $\Phi(g\mathbf{z}) = \varphi_j(g)\tilde{Z}$, $SU(2)$-equivariance of B is seen as follows:

$$B_{P^g}(\mathbf{n}) = h(\tilde{Z}, P^g\tilde{Z}) = h(\tilde{Z}, \varphi_j(g)P\varphi_j(g)^{-1}\tilde{Z}) = h(\varphi_j(g)^{-1}\tilde{Z}, P\varphi_j(g)^{-1}\tilde{Z})$$

$$= h(\Phi(g^{-1}\mathbf{z}), P\Phi(g^{-1}\mathbf{z})) = B_P(\pi(g^{-1}\mathbf{z})) = B_P(g^{-1}\mathbf{n}) = (B_P)^g(\mathbf{n}).$$

$\square$

Lemma 6.2.29. *The map $B : P \mapsto B_P$ satisfies the reality condition.*

Proof. By the definition of B_P,

$$B_P(\mathbf{n}) = h(\tilde{Z}, P\tilde{Z}) = \sum_{i=0}^{n}\sum_{j=0}^{n} b_i b_j \, p_{ji} z_1^{n-i} z_2^i \overline{z_1}^{n-j} \overline{z_2}^j$$

$$= \sum_i b_i^2 \, p_{ii} |z_1|^{2(n-i)}|z_2|^{2i} + \sum_{j<i} b_i b_j |z_1|^{2(n-i)}|z_2|^{2j}[p_{ij}(\overline{z_1}z_2)^{i-j} + p_{ji}(z_1\overline{z_2})^{i-j}],$$

where we have written $b_i = \sqrt{\binom{n}{i}}$ for simplicity. Now, the replacement $P \to P^*$ means $p_{ii} \to \overline{p_{ii}}$ and

$$[p_{ij}(\overline{z_1}z_2)^{i-j} + p_{ji}(z_1\overline{z_2})^{i-j}] \to [\overline{p_{ij}(\overline{z_1}z_2)^{i-j} + p_{ji}(z_1\overline{z_2})^{i-j}}],$$

and consequently $B_{P^*}(\mathbf{n}) = \overline{B_P(\mathbf{n})}$. $\square$

Lemma 6.2.30. *The map $B : P \mapsto B_P$ is injective.*

Proof. The kernel of B is $Ker(B) = \mathcal{K} = \{P \in M_{\mathbb{C}}(n+1); B_P = 0\}$, which is an $SU(2)$-invariant subspace of $M_{\mathbb{C}}(n+1) = \sum_{l=0}^{n} M_{\mathbb{C}}(\varphi_l)$, so assuming $\mathcal{K} \neq 0$ it splits as a direct sum of irreducible subspaces

$$Ker(B) = \mathcal{K} = \sum_{i=1}^{k} \mathcal{K}(\varphi_{l_i}),$$

then each of the summands has a zero weight vector $|l_i 0 >$, namely there is some non-zero diagonal matrix

$$D = (d_0, d_1, \ldots, d_n)_0 \in \mathcal{K}$$

and consequently, for each $\mathbf{n} \in S^2$ there is the linear equation

$$B_D(\mathbf{n}) = h(\tilde{Z}, D\tilde{Z}) = \sum_{i=0}^{n} d_i b_i^2 z_1^{n-i} z_2^i \overline{z_1}^{n-i} \overline{z_2}^i = \sum_{i=0}^{n} d_i b_i^2 |z_1|^{2(n-i)} |z_2|^{2i} = 0$$

for the "variables" d_i. Clearly, the only common solution is $d_i = 0$ for each i, so we must have $\mathcal{K} = 0$.   □

Lemma 6.2.31. *The map B can be expressed by the formula*

$$B_Q(g\mathbf{n}_0) = trace(Q\Pi_1^g), \ g \in SU(2), \tag{6.34}$$

where $\mathbf{n}_0 = (0, 0, 1) \in S^2$ *is the north pole, and* Π_1 *is the projection operator*

$$\Pi_1 = diag(1, 0, 0, \ldots, 0).$$

Proof. Referring to (6.31) and (6.32), we have

$$\Phi(1, 0) = \tilde{Z}_0 = (1, 0, \ldots, 0) \text{ and } \pi(1, 0) = \mathbf{n}_0,$$

consequently for $Q = (q_{ij})$, $1 \le i, j \le n + 1$,

$$B_Q(\mathbf{n}_0) = h(\tilde{Z}_0, Q\tilde{Z}_0) = q_{11} = trace(Q\Pi_1).$$

Therefore (6.34) holds for $g = e$, and hence by the equivariance of B, $B_Q(g\mathbf{n}_0) = B_{Q^{g^{-1}}}(\mathbf{n}_0) = trace(Q^{g^{-1}}\Pi_1) = trace(Q\Pi_1^g)$ □

Corollary 6.2.32. *The map B is the standard Berezin symbol correspondence (cf. Definition 6.2.12).*

Remark 6.2.33. Berezin [12] introduced his symbol correspondence in a manner similar to, but not equal to the one presented here. Instead of using the space of homogeneous polynomials in two variables ${}^h P_\mathbb{C}^n(z_1, z_2)$, Berezin used the representation on the space of holomorphic polynomials on the sphere $\mathcal{H}ol^n(S^2)$. As the two representations are intimately connected (see the discussion at the end of Sect. 2.2), it is not hard to see that the symbols obtained by Eq. (6.33) coincide with Berezin's original definition of covariant symbols on the sphere.

Now, the characteristic numbers of the standard Berezin symbol correspondence are the characteristic numbers of the symbol map whose operator kernel K is the projection operator Π_1. Thus for each $n = 2j$ there is a string $\vec{b}$ of n characteristic numbers possibly depending on n, which in this case shall be denoted by b_l^n, namely

$$\vec{b} = (b_1^n, b_2^n, \ldots, b_l^n, \ldots, b_n^n).$$

According to (6.16), these numbers are expressed by the inner product

$$b_l^n = \sqrt{\frac{n+1}{2l+1}} \left\langle \Pi_1, \mathbf{e}^j(l,0) \right\rangle = \sqrt{\frac{n+1}{2l+1}} \mathbf{e}^j(l,0)_{1,1}, \tag{6.35}$$

where $\mathbf{e}^j(l,0)_{1,1}$ denotes the first entry of the diagonal matrix $\mathbf{e}^j(l,0)$.

From Remark 3.3.11, Eqs. (3.83) and (3.84), we have that

$$b_l^n = \sqrt{\frac{n+1}{2l+1}} C_{j,-j,0}^{j,j,l}. \tag{6.36}$$

Thus, the explicit expression for b_l^n can be obtained from the general formulae for Clebsh-Gordan coefficients, as (3.47). But it can also be obtained directly by induction, as shown in Appendix 10.5. The result is expressed below.

Proposition 6.2.34. *The characteristic numbers b_l^n of the standard Berezin symbol correspondence are given explicitly by*

$$b_l^n = \sqrt{\frac{n(n-1)\ldots(n-l+1)}{(n+2)(n+3)\ldots(n+l+1)}} = \frac{\sqrt{\binom{n}{l}}}{\sqrt{\binom{n+l+1}{l}}} = \frac{n!\sqrt{n+1}}{\sqrt{(n+l+1)!(n-l)!}}. \tag{6.37}$$

Corollary 6.2.35. *The standard Berezin symbol correspondence is characteristic-positive and, $\forall l \in \mathbb{N}$, its characteristic numbers b_l^n expand in powers of n^{-1} as:*

$$b_l^n = 1 - \frac{1}{n}\binom{l+1}{2} + \frac{1}{n^2}\left\{ \frac{1}{2}\binom{l+1}{2}^2 + \binom{l+1}{2} \right\}$$

$$- \frac{1}{n^3}\left\{ \frac{1}{6}\binom{l+1}{2}^3 + \frac{4}{3}\binom{l+1}{2}^2 + \binom{l+1}{2} \right\} + \ldots \tag{6.38}$$

According to Summary 6.2.17, the standard Berezin symbol correspondence, with operator kernel being the projection operator Π_1, can be continuously deformed to the standard Stratonovich-Weyl symbol correspondence. On the other hand, according to Remark 6.2.11, $\forall n \geq 1$, the projection operator Π_{n+1} is also the operator kernel of a Berezin symbol correspondence. Can it be defined via the Hermitian metric h? What are its characteristic numbers?

Proposition 6.2.36. *Let* $\sigma : \mathbb{C}^2 \to \mathbb{C}^2$ *be given by*

$$\sigma : \mathbf{z} = (z_1, z_2) \mapsto (-\bar{z}_2, \bar{z}_1)$$

and let $\Phi^- = \Phi \circ \sigma : \mathbb{C}^2 \to \mathbb{C}^{n+1}$, *for* Φ *as in (6.31). The map*

$$B^- : M_{\mathbb{C}}(n+1) \to C^\infty(S^2) , \; P \mapsto B_P^- ,$$

defined by

$$B_P^-(\mathbf{n}) = h(\Phi^-(\mathbf{z}), P\Phi^-(\mathbf{z})) , \qquad (6.39)$$

is a symbol correspondence with operator kernel Π_{n+1} *and characteristic numbers*

$$b_{l_-}^n = (-1)^l b_l^n , \qquad (6.40)$$

for b_l^n *given by (6.37).*

Proof. We note that, under the Hopf map $\pi : S^3 \subset \mathbb{C}^2 \to S^2$ given by (2.31), the conjugate of σ, given by $\pi \circ \sigma = \alpha \circ \pi$, is the antipodal map $\alpha : S^2 \to S^2$,

$$\alpha : \mathbf{n} \mapsto -\mathbf{n} .$$

It follows that the function B_P^- given by (6.39) satisfies $B_P^- = B_P \circ \alpha$, that is,

$$\forall P \in M_{\mathbb{C}}(n+1), \; B_P^-(\mathbf{n}) = B_P(-\mathbf{n}) , \qquad (6.41)$$

where B_P is the standard Berezin symbol of P. Thus, clearly, all properties of Definition 6.1.1 are satisfied for the map B^-.

Now, if $P = \mu_0 e^j(l, m)$, then, according to (6.12), $B_P^-(\mathbf{n}) = b_{l_-}^n Y_l^m(\mathbf{n})$. On the other hand, by (6.12) and (4.31), $B_P(-\mathbf{n}) = b_l^n Y_l^m(-\mathbf{n}) = (-1)^l b_l^n Y_l^m(\mathbf{n})$. Therefore, (6.40) follows from (6.41).

Finally,

$$B_P^-(\mathbf{n}_0) = h(P\Phi^-(1,0), \Phi^-(1,0))$$

$$= h(P(0, 0, \ldots, 0, 1), (0, 0, \ldots, 0, 1))$$

$$= p_{(n+1)(n+1)},$$

for $P = [p_{ij}]$. Thus,

$$B_P^-(\mathbf{n}_0) = trace(P\Pi_{n+1})$$

and, by equivariance, Π_{n+1} is the operator kernel for B^-.   □

Definition 6.2.37. The symbol correspondence defined by the characteristic numbers $b_{l-}^n = (-1)^l b_l^n$ is called the *alternate Berezin symbol correspondence*.

According to Summary 6.2.17, the alternate Berezin symbol correspondence can be continuously deformed to a Stratonovich-Weyl symbol correspondence:

Definition 6.2.38. The unique Stratonovich-Weyl symbol correspondence continuously deformed from the alternate Berezin symbol correspondence is called the *alternate Stratonovich-Weyl symbol correspondence* and is given by the characteristic numbers $c_l^n \equiv \varepsilon_l = (-1)^l$. It shall be denoted by

$$W_{1-}^j : \mathcal{B}(\mathcal{H}_j) \simeq M_{\mathbb{C}}(n+1) \to Poly_{\mathbb{C}}(S^2)_{\leq n} \subset C_{\mathbb{C}}^{\infty}(S^2).$$

Definition 6.2.39. If $W_{\vec{c}}^j$ is any characteristic-positive symbol correspondence given by characteristic numbers $c_l^n > 0$, the symbol correspondence $W_{\vec{c}-}^j$ with characteristic numbers $c_{l-}^n = (-1)^l c_l^n$ is a *characteristic-alternate* symbol correspondence. More generally, we say that two symbol correspondences defined by characteristic numbers $\vec{c} = (c_1^n, \cdots, c_n^n)$ and $\vec{c}' = (c_1^{n'}, \cdots, c_n^{n'})$ stand in *alternate-relation*, or *antipodal-relation* to each other if $c_l^{n'} = (-1)^l c_l^n$, $\forall l \leq n$.

6.2.4 Symbol Correspondences via Coherent States

Although the Stratonovich-Weyl symbol correspondences satisfy all properties (i)–(iv) of general symbol correspondences plus the isometry property (v) and therefore seem to be preferable or "better" correspondences, they all seemingly fail to satisfy a property that is satisfied by all Berezin correspondences.

To look into this, let us first recall and state some basic definitions.

Definition 6.2.40. A Hermitian operator $M \in M_{\mathbb{C}}(n+1)$ is *positive* (resp. *positive-definite*), if all of its eigenvalues are nonnegative (resp. positive).[1] A real function $f \in C_{\mathbb{C}}(S^2)$ is *positive* (resp. *strictly-positive*) if $f(\mathbf{n}) \geq 0$ (resp. $f(\mathbf{n}) > 0$), $\forall \mathbf{n} \in S^2$. Then, a pre-symbol map $M_{\mathbb{C}}(n+1) \to C_{\mathbb{C}}(S^2)$ is a *positive map* if it takes positive(-definite) operators to (strictly-)positive functions.

Definition 6.2.41. A symbol correspondence $W_{\vec{c}}^j : M_{\mathbb{C}}(n+1) \to Poly_{\mathbb{C}}(S^2)_{\leq n}$ is *mapping-positive* if it is a positive map.

Now, we know that

$$span_{\mathbb{R}}\{I, \mathbf{e}(l, 0)\}_{1 \leq l \leq n} = span_{\mathbb{R}}\{\Pi_k\}_{1 \leq k \leq n+1} = \Delta_{\mathbb{R}}(n+1)$$

---

[1] This definition can be generalized for a non-Hermitian operator by looking at its Hermitian factor, but we'll stick here to the simpler definition which requires the operator to be Hermitian.

is the space of all real diagonal matrices inside $M_{\mathbb{C}}(n+1)$. A subset of $\Delta_{\mathbb{R}}(n+1) \simeq \mathbb{R}^{n+1}$ is the set (affine hyperplane) of all trace-one real diagonal matrices

$$\Delta_{\mathbb{R}}^1(n+1) = \frac{1}{n+1} I \oplus span_{\mathbb{R}}\{\mathbf{e}(l,0)\}_{1 \le l \le n} \subset \Delta_{\mathbb{R}}(n+1),$$

which via (6.13) is isomorphic to the space of pre-symbol maps $M_{\mathbb{C}}(n+1) \to C_{\mathbb{C}}(S^2)$. Then, a subset of $\Delta_{\mathbb{R}}^1(n+1) \simeq \mathbb{R}^n$ is the set of operator kernels of a symbol correspondence, which is defined via (6.13), so let us denote this subset by

$$\mathcal{W}[j] \subset \Delta_{\mathbb{R}}^1(n+1)$$

and we know from Theorem 6.1.7 that $\mathcal{W}[j] \simeq (\mathbb{R}^*)^n$ (we recall that a pre-symbol map is a symbol correspondence if it is injective, cf. Definition 6.1.1).

   If W^{k_1} and $W^{k_2} : M_{\mathbb{C}}(n+1) \to Poly_{\mathbb{C}}(S^2)_{\le n}$ are two Berezin correspondences defined respectively by projectors Π_{k_1} and $\Pi_{k_2} \in \mathcal{W}[j] \subset \Delta_{\mathbb{R}}^1(n+1)$, it follows from (6.13) that $K = a\Pi_{k_1} + (1-a)\Pi_{k_2} \subset \Delta_{\mathbb{R}}^1(n+1)$ is the operator kernel of a symbol correspondence if the characteristic numbers of K are all non-zero.

   More specifically, by taking convex combinations $K = a\Pi_{k_1} + (1-a)\Pi_{k_2}$, $a \in [0,1]$, all characteristic numbers of K are convex combinations of the characteristic numbers of Π_{k_1} and Π_{k_2} with the same $a \in [0,1]$ and, for at most n values of a, at least one of the characteristic numbers of K vanishes. For the remaining values of a (a dense subset of $[0,1]$), K defines a symbol correspondence via (6.13).

   Following standard terminology in quantum mechanics, we have:

Definition 6.2.42. If $\Pi_{\vec{v}}$ is the projector on the 1-dimensional subspace generated by $\vec{v} \in \mathcal{H}_j \simeq \mathbb{C}^{n+1}$, then $\Pi_{\vec{v}}$ is a *pure state*. Given two pure states, $\Pi_{\vec{v}_1}, \Pi_{\vec{v}_2}$, any convex combination $S_a = a\Pi_{\vec{v}_1} + (1-a)\Pi_{\vec{v}_2}$, $a \in [0,1]$, defines a *mixed state* if $a \ne 0, 1$. More generally, $S \in \mathcal{B}(\mathcal{H}_j) \simeq M_{\mathbb{C}}(n+1)$ is a *generalized state*, or simply a *state*, if S is any convex combination of any number of pure states.

Proposition 6.2.43. *An operator $S \in \mathcal{B}(\mathcal{H}_j) \simeq M_{\mathbb{C}}(n+1)$ is a state if and only if S is a positive Hermitian operator of trace 1.*

Proof. The "only if" direction is immediate from Definition 6.2.42. On the other hand, if S is a positive Hermitian operator of trace 1, then it can be diagonalized and all its eigenvalues a_k satisfy $a_k \ge 0$ and furthermore $a_1 + a_2 + \cdots + a_{n+1} = 1$. Hence, S is a convex combination of its eigen-projectors. $\square$

   Thus, for the standard spin-j basis, let

$$Conv\{\Pi_k\}_{n+1} \subset \Delta_{\mathbb{R}}^1(n+1)$$

denote the convex hull of all 1-dimensional projectors Π_k, that is, $P \in Conv\{\Pi_k\}_{n+1}$ iff $P = a_1\Pi_1 + a_2\Pi_2 + \cdots + a_{n+1}\Pi_{n+1}$, s.t. $a_1 + a_2 + \cdots a_{n+1} = 1$, $\mathbb{R} \ni a_k \ge 0$, $\forall k$.

Then, $Conv\{\Pi_k\}_{n+1} \subset \Delta_{\mathbb{R}}^1(n+1)$ defines the set of all states which are fixed by the 1-torus, or circle group $U(1) = \{\exp(itJ_3), \ t \in \mathbb{R}\}$, and therefore defines the set of all J_3-invariant states. Then we have the following:

Theorem 6.2.44. *An operator kernel $K \in \mathcal{W}[j]$ defines a mapping-positive symbol correspondence $W^j : M_{\mathbb{C}}(n+1) \rightarrow Poly_{\mathbb{C}}(S^2)_{\leq n}$ via Eq. (6.13) if and only if K is a J_3-invariant state, that is, $K \in \mathcal{W}[j] \cap Conv\{\Pi_k\}_{n+1}$.*

Proof. First, recall from linear algebra that a Hermitian operator $M \in M_{\mathbb{C}}(n+1)$ is positive iff $M = S^*S$, for some $S \in M_{\mathbb{C}}(n+1)$, and $M = S^*S$ is positive-definite iff $\det(S) \neq 0$ (in this latter case, there is a unique upper triangular matrix S with positive diagonal elements and this factorization of M is called the Cholesky decomposition).

Now, if Π is the projector onto the 1-dimensional subspace spanned by a unitary vector $\vec{v} \in \mathbb{C}^{n+1}$, then $\Pi = \Pi_{\vec{v}} = (\vec{v})(\vec{v})^*$, where $(\vec{v})$ is the $(n+1)$-column vector representing $\vec{v}$ in a certain basis. It follows that

$$trace(\Pi_{\vec{v}} M) = trace((\vec{v})(\vec{v})^* S^* S) = trace(S(\vec{v})(S(\vec{v}))^*) = |S\vec{v}|^2 \geq 0 , \tag{6.42}$$

with strict inequality if S is non-singular, i.e. M is positive-definite.

Then, from (6.13) and (6.42) we have that

$$trace(\Pi_k^g M) = trace(\varphi_j(g)\Pi_k\varphi_j(g)^* S^* S) = trace(S\varphi_j(g)\Pi_k\varphi_j(g)^* S^*)$$
$$= trace(T \Pi_k T^*) = trace(\Pi_k T^* T) \geq 0, \tag{6.43}$$

where $T = S\varphi_j(g)$ and, again, strict inequality follows if $\det(T) = \det(S) \neq 0$.

Since (6.43) holds for all $g \in G = SU(2)$, it follows that the pre-symbol map

$$B^{j,k} : M_{\mathbb{C}}(n+1) \rightarrow \mathcal{C}_{\mathbb{C}}(S^2) , \quad M \mapsto trace(\Pi_k^g M), g \in G,$$

is a positive map, that is, $B^{j,k}(M) = B_M^{j,k} \in \mathcal{C}_{\mathbb{C}}(S^2)$ is a positive function if M is a positive operator. And it follows immediately that if $K \in Conv\{\Pi_k\}_{n+1}$, then the pre-symbol map $M \mapsto trace(K^g M), g \in G$, is a positive map.

Therefore, this pre-symbol map is a mapping-positive symbol correspondence if all characteristic numbers are non-zero.

We have proved that if $K \in \mathcal{W}[j] \cap Conv\{\Pi_k\}_{n+1}$, then K defines a mapping-positive symbol correspondence via (6.13).

On the other hand, if $K \in \mathcal{W}[j] \setminus Conv\{\Pi_k\}_{n+1}$, then $K = a_1\Pi_1 + a_2\Pi_2 + \cdots + a_{n+1}\Pi_{n+1}$ such that $a_1 + a_2 + \cdots a_{n+1} = 1$, but at least one $a_k < 0$. Then, if M is a positive diagonal matrix whose only positive eigenvalue is the kth eigenvalue, it follows that $trace(KM) < 0$ and therefore K does not define a mapping-positive symbol correspondence via (6.13). $\square$

Remark 6.2.45. We recall from Definition 3.1.4 that a standard basis of $\mathcal{H}_j$ is a set of eigenstates of $J_{\mathbf{n}_0} = J_3$ and, since this universal convention of singling out J_3 is used throughout the whole book, one might wonder why we are emphasizing in the statement of Theorem 6.2.44 that these states are J_3-invariant states. The reason will become clear a bit further below, after Definition 6.2.50.

The following corollary of Theorem 6.2.44 is immediate:

Corollary 6.2.46. *Every Berezin spin-j symbol correspondence is a mapping-positive symbol correspondence.*

Remark 6.2.47. On the other hand, clearly not all mapping-positive spin-j symbol correspondences are of Berezin type. Although there are at most $n + 1$ Berezin spin-j symbol correspondences, clearly the set $\mathcal{W}[j] \cap Conv\{\Pi_k\}_{n+1}$ is dense in $Conv\{\Pi_k\}_{n+1}$, which has positive measure in the space of pre-symbol maps $\Delta_{\mathbb{R}}^1(n + 1) \simeq \mathbb{R}^n$ (for its standard measure, say), so the set of all mapping-positive symbol correspondences has positive measure in the moduli-space $(\mathbb{R}^*)^n$ of all spin-j symbol correspondences (with its induced measure from $\mathbb{R}^n$).

However, we seem to have the following:

Conjecture 6.2.48. No Stratonovich-Weyl spin-j symbol correspondence is a mapping-positive symbol correspondence.

We have verified this conjecture for small j, namely in the range $1/2 \le j \le 3$, but we leave the general case as an open problem. Here we show the argument for $j \le 3/2$, using the matrices exhibited in Examples 3.3.8–3.3.10.

Thus, we must check that if K is the J_3-invariant operator kernel of a Stratonovich-Weyl symbol correspondence in the sense of (6.13), then K is not a J_3-invariant state, i.e., $K \notin Conv\{\Pi_k\}_{n+1}$.

We start with $j = 1/2$, i.e. $n = 1$. In this case, K is the kernel of a Stratonovich-Weyl symbol correspondence iff $K = \frac{1}{2}(I \pm \sqrt{3}\sigma_3)$. On the other hand, $K \in Conv\{\Pi_k\}_2$ iff $K = \frac{1}{2}(I + \eta\sigma_3)$, $\eta \in [-1, 1]$. But clearly, $\pm\sqrt{3} \notin [-1, 1]$.

We now look at $j = 1$, i.e. $n = 2$. From Example 3.3.9 and (6.16) we have that K is the operator kernel of a Stratonovich-Weyl symbol correspondence iff

$$K = \left(\frac{\sqrt{2} + 3\varepsilon_1 + \sqrt{5}\varepsilon_2}{3\sqrt{2}}\right)\Pi_1 + \left(\frac{1 - \sqrt{10}\varepsilon_2}{3}\right)\Pi_2 + \left(\frac{\sqrt{2} - 3\varepsilon_1 + \sqrt{5}\varepsilon_2}{3\sqrt{2}}\right)\Pi_3 \,,$$

where $\varepsilon_l = \pm 1$. But if $a_1\Pi_1 + a_2\Pi_2 + a_3\Pi_3 \in Conv\{\Pi_k\}_3$, then necessarily every $a_k \in [0, 1]$. But clearly, $\frac{1 - \sqrt{10}\varepsilon_2}{3} \notin [0, 1]$.

Similarly for the case $j = 3/2$, i.e. $n = 3$, from Example 3.3.10 and (6.16), K is the operator kernel of a Stratonovich-Weyl symbol correspondence iff

$$K = \left(\frac{\sqrt{5} + 3\sqrt{3}\varepsilon_1 + 5\varepsilon_2 + \sqrt{7}\varepsilon_3}{4\sqrt{5}} \right) \Pi_1 + \left(\frac{\sqrt{5} + \sqrt{3}\varepsilon_1 - 5\varepsilon_2 - 3\sqrt{7}\varepsilon_3}{4\sqrt{5}} \right) \Pi_2$$

$$+ \left(\frac{\sqrt{5} - \sqrt{3}\varepsilon_1 - 5\varepsilon_2 + 3\sqrt{7}\varepsilon_3}{4\sqrt{5}} \right) \Pi_3 + \left(\frac{\sqrt{5} - 3\sqrt{3}\varepsilon_1 + 5\varepsilon_2 - \sqrt{7}\varepsilon_3}{4\sqrt{5}} \right) \Pi_4$$

where again $\varepsilon_l = \pm 1$. But again, $K \in Conv\{\Pi_k\}_4$ only if all coefficients a_k of Π_k lie in $[0, 1]$, which one can verify not to be the case for any choice of ε_l's.

We have checked the validity of this conjecture up to $j = 3$, i.e., $n = 6$, but no further. Thus, one way to prove (or disprove) the conjecture for the general case, is to show that for every $n \in \mathbb{N}$ there is no choice of $\varepsilon_l = \pm 1$, $1 \leq l \leq n = 2j$, (or there is at least one such choice) such that all $a_k \in [0, 1]$, $1 \leq k \leq n + 1$, where

$$a_k = \frac{1}{n+1} + (-1)^{k+1} \sum_{l=1}^{n} \varepsilon_l \sqrt{\frac{2l+1}{n+1}} C_{m,-m,0}^{j,\,j,\,l}, \quad m = j - k + 1. \qquad (6.44)$$

Remark 6.2.49. For affine mechanical systems, a statement partially analogous to Conjecture 6.2.48 is well-known to be true. Namely, the Weyl symbol correspondence, which satisfies a similar isometry condition (cf. (5.10)) and is the analogue of the standard Stratonovich-Weyl symbol correspondence, also does not satisfy a similar positivity condition. In that context, the Weyl symbol of a pure state ψ is called the Wigner function of ψ and the very well-known fact that the Wigner function of a pure state takes on negative values has important consequences.

Hence, assuming the validity of Conjecture 6.2.48, the set of isometric spin-j symbol correspondences is disjoint from the set of mapping-positive ones. This latter property (mapping-positivity) often plays a helpful role in some analytical investigations of the symbols, while the isometry property simplifies and clarifies matters considerably in other aspects, so we have a competition between these two kinds of spin-j symbol correspondences and none seems to possess both properties.

For the sake of further clarity, it is useful to recall some other standard terminology from quantum mechanics:

Definition 6.2.50. If S is a J_3-invariant state, that is $S \in Conv\{\Pi_k\}_{n+1}$, let

$$S^g = \varphi_j(g) S \varphi_j(g)^{-1} = S(\mathbf{n}), \quad \mathbf{n} = g\mathbf{n}_0. \qquad (6.45)$$

Then, $S^g = S(\mathbf{n})$ is a *coherent state*.

More precisely, $S^g = S(\mathbf{n})$ is actually a *family of states parametrized by the points on the 2-sphere*, such that for each $\mathbf{n} \in S^2$, $S^g = S(\mathbf{n})$ is a state which is fixed by the circle group $U(1) = \{\exp(itJ(\mathbf{n})), \ t \in \mathbb{R}\}$, where

$$J(\mathbf{n}) = J_3^g = \varphi_j(g) J_3 \varphi_j(g)^{-1}, \quad \mathbf{n} = g\mathbf{n}_0. \qquad (6.46)$$

The point being that, as **n** varies along the sphere, the state $S(\mathbf{n})$ also varies in a "coherent" way, that is, equivariantly with respect to the $SU(2)$ group action.[2]

Lemma 6.2.51. *The operator-valued function on S^2 defined in (6.46) is "the component of total angular momentum* **J** *in the direction* **n**", *that is,* $J(\mathbf{n}) = J_{\mathbf{n}} = \mathbf{n} \cdot \mathbf{J}$.

Proof. Both functions agree at the north pole, $J(\mathbf{n}_0) = J_{\mathbf{n}_0} = J_3$, and are defined as functions on S^2 via left action of $SO(3)$.   □

Definition 6.2.52. If S is a state (cf. Definition 6.2.42 and Proposition 6.2.43) and $P \in \mathcal{B}(\mathcal{H}_j)$ is an operator, then $trace(SP)$ is the *expectation value of the operator P in the state S*, or simply the *expectation of P in S*.

It follows that every mapping-positive symbol correspondence is defined as the set of expectation values of operators in a certain coherent (family of) state(s).

That is, if W^j is the symbol correspondence defined by the operator kernel $K = S \in Conv\{\Pi_k\}_{n+1} \cap \mathcal{W}[j]$ via (6.13), then for any operator $P \in M_{\mathbb{C}}(n+1)$, its symbol W_P^j is the function whose value at any point $\mathbf{n} \in S^2$ is the expectation value of P in the state $S(\mathbf{n}) = S^g$.

For the standard Berezin correspondence, $S_>(\mathbf{n}) = S_>^g = \Pi_1^g$ is called the *highest-weight* or *standard* coherent (family of) state(s), cf. Remark 6.2.13. For the alternate case, $S_<(\mathbf{n}) = S_<^g = \Pi_{n+1}^g$ is the *lowest-weight* coherent state.

Thus, one important subset of general spin-j symbol correspondences is the set of all symbol correspondences defined by coherent states (the mapping-positive ones) and another important subset of general symbol correspondences is defined by the isometric ones. But assuming Conjecture 6.2.48, these two subsets are disjoint.

Now, in view of the above discussion, it would be interesting to have other explicit examples of spin-j symbol correspondences that are defined by coherent states. To this end, let us start from the pre-symbol maps defined by projectors, and denoting the characteristic numbers of the standard and alternate Berezin correspondences by

$$b_l^n = b_l^n(1) \,, \; b_{l-}^n = b_l^n(n+1) \,,$$

we have from (6.16) and (6.18) that

$$b_l^n(k) = (-1)^{k+1} \sqrt{\frac{n+1}{2l+1}} C_{m,-m,0}^{j,\,j,\,l} \,, \; m = j - k + 1 \,, \tag{6.47}$$

are the characteristic numbers of a Berezin correspondence defined via (6.13) by $K = \Pi_k$, *provided that* $b_l^n(k) \neq 0$, $\forall l = 1, 2, \cdots, n$. So, the decisive question is whether the Clebsch-Gordan coefficients in (6.47) vanish or not.

---

[2]Although in the literature it is more common to find the term "coherent state" referring to a coherent family of pure states, for our purposes such a strict restriction is not appropriate.

However, even though these coefficients are special cases, closed formulae for $C_{m,-m,0}^{j,j,l}$ are not known for every m. But irrespective of the latter non-vanishing condition, we will still refer to $b_l^n(k)$ as the characteristic numbers of the Berezin pre-symbol map defined by $K = \Pi_k$ via (6.13).

Let us consider some cases of (6.47). First note that m lies in the range $-j \leq m \leq j$, and the extreme cases $m = j$ and $m = -j$ refer to the well-known standard and alternate Berezin cases, $k = 1$ and $k = n + 1$, respectively, which as mentioned above, refer to the highest-weight and lowest-weight coherent states. Therefore, these symbol correspondences could also be called the *highest-state* and the *lowest-state* spin-j symbol correspondences.

From the cases $k = 1$ and $k = n + 1$ one can easily determine the coefficients for $k = 2$ and $k = n$ using some recursive formulas (cf. [74]), so we get for $n > 2$,

$$b_l^n(2) = \left(1 - \frac{l(l+1)}{n}\right) b_l^n \ , \ b_l^n(n) = \left(1 - \frac{l(l+1)}{n}\right) b_{l-}^n \ . \tag{6.48}$$

Consequently, both Π_2 and Π_n fail to be the operator kernel of a symbol correspondence whenever $n = 2j$ can be expressed as the product $n = l(l+1) > 2$ for some integer l, namely $n = 6, 12, 20, 30, \ldots$ (cf. Remark 6.2.11, where these obstructions were stated without explicit formulas for the characteristic numbers and it is observed that, at least for $n \leq 20$, no $b_l^n(k)$ vanishes when n is odd).

To investigate other cases, it follows from the symmetry properties of the Clebsch-Gordan coefficients (cf. (3.50)) and (6.47) that we have the general relation

$$b_l^n(k) = (-1)^l b_l^n(n + 2 - k) \ , \tag{6.49}$$

that is, $b_l^n(k)$ and $b_l^n(n + 2 - k)$ stand in alternate relation to each other.

Therefore, it suffices to look at the cases $k \leq [j + 1]$, where $[j + 1]$ denotes the integral part of $j + 1$. So, let us now look at the other extremal possibility from $k = 1$, in other words, let us look for a symbol correspondence which is defined by a coherent state that is as far as possible from the highest coherent state, while still within the convex hull of the "first half" of projectors.

Consider first the case $k = j + 1$, so j is integral and $m = 0$ in (6.47). The Clebsch-Gordan coefficients $C_{0,0,0}^{j,j,l}$ are described in Remark 4.2.7 (cf. (4.49)); in particular, $C_{0,0,0}^{j,j,l}$ is zero if and only if $2j + l$ is odd, that is, l is odd. In other words, when j is an integer, the pre-symbol map defined by $K = \Pi_{j+1}$ via (6.13) is very far from defining a symbol correspondence. Nonetheless, we shall see that Π_{j+1} and Π_j can be combined to define a symbol correspondence via (6.13), whereas for j half-integral, $\Pi_{j+1/2}$ does, in fact, define a symbol correspondence.

Thus, noting that $\Pi_{[j+1]} = \Pi_{j+1}$ when j is integer and $\Pi_{[j+1]} = \Pi_{j+1/2}$ when j is half-integer, but that in the latter case we can write $\Pi_{j+1/2} = \Pi_{[j+1/2]}$, while in the integral j case we have $\Pi_{[j+1/2]} = \Pi_j$, we define:

Definition 6.2.53. For every $n = 2j \in \mathbb{N}$, the *upper-middle coherent state* is the coherent (family of) state(s) $S_{1/2}(\mathbf{n}) = S_{1/2}^g$, where $S_{1/2}(\mathbf{n}_0) = S_{1/2}$ is given by

$$S_{1/2} = \frac{1}{2}\left(\Pi_{[j+1/2]} + \Pi_{[j+1]}\right).$$ (6.50)

Thus we note that $S_{1/2}$ is a pure state when j is half-integer, but it is a mixed state when j is integer. The notation $S_{1/2}$ comes from the fact that the magnetic moment m for this state is $m = 1/2$ when j is half-integer, while in the integral j case the value $1/2$ is the average (expected) value of J_3 in this state, that is

$$trace(S_{1/2}J_3) = trace(S_{1/2}(\mathbf{n})J(\mathbf{n})) = 1/2, \ \forall \mathbf{n} \in S^2, \ \forall n = 2j \in \mathbb{N}.$$ (6.51)

On the other hand, clearly the value $1/2$ will not generally be the expectation value of $J(\mathbf{n})$ in the state $S_{1/2} = S_{1/2}(\mathbf{n}_0)$, if $\mathbf{n} \neq \mathbf{n}_0$, or equivalently, $1/2$ will not generally be the expectation value of $J_3 = J(\mathbf{n}_0)$ in the state $S_{1/2}(\mathbf{n})$, if $\mathbf{n} \neq \mathbf{n}_0$.

Proposition 6.2.54. *For every $n = 2j \in \mathbb{N}$, the operator kernel $K = S_{1/2}$ given by (6.50) defines a spin-j symbol correspondence via (6.13), whose characteristic numbers c_l^n, henceforth denoted $p_l^n(1/2)$, are given below.*

When $n = 2k - 1$, $l = 2q$, $p_l^n(1/2) = p_{2q}^{2k-1}(1/2) =$

$$(-1)^q \frac{(2q)!}{(q!)^2} \sqrt{\frac{2}{k}} \frac{(k+q)(k+q-1)\cdots(k-q)}{\sqrt{(2k+2q)(2k+2q-1)\cdots(2k-2q)}}.$$ (6.52)

When $n = 2k - 1$, $l = 2q + 1$, $p_l^n(1/2) = p_{2q+1}^{2k-1}(1/2) =$

$$(-1)^q \frac{(2q+1)!}{(q!)^2} \sqrt{\frac{2}{k}} \frac{(k+q)(k+q-1)\cdots(k-q)}{\sqrt{(2k+2q+1)(2k+2q)\cdots(2k-2q-1)}}.$$ (6.53)

When $n = 2k$, $l = 2q$, $p_l^n(1/2) = p_{2q}^{2k}(1/2) =$ (6.54)

$$(-1)^q \frac{(2q)!}{(q!)^2} \left(1 - \frac{q(2q+1)}{k(2k+2)}\right) \frac{\sqrt{2k+1}(k+q)(k+q-1)\cdots(k-q+1)}{\sqrt{(2k+2q+1)(2k+2q)\cdots(2k-2q+1)}}.$$

When $n = 2k$, $l = 2q + 1$, $p_l^n(1/2) = p_{2q+1}^{2k}(1/2) =$

$$(-1)^q \frac{(2q+1)!}{(q!)^2} \frac{\sqrt{(2k+1)(2k+2q+2)}(k+q)(k+q-1)\cdots(k-q)}{2k(k+1)\sqrt{(2k+2q+1)(2k+2q)\cdots(2k-2q)}}.$$

(6.55)

Definition 6.2.55. The spin-j symbol correspondence defined by the operator kernel $K = S_{1/2}$ shall be called the *upper-middle-state* symbol correspondence.

The proof of Proposition 6.2.54 is found in Appendix 10.6.

Remark 6.2.56. While the projector $\Pi_{[j+1]}$ fails miserably to define a symbol correspondence when j is any integer, the projector $\Pi_{[j+1/2]}$ fails to generally define a symbol correspondence by not much. Apart from defining a symbol correspondence for all odd values of n, it only fails to define a symbol correspondence for even values of $n = 2j$ when $j(j + 1) = 2p(4p + 1)$, for some $p \in \mathbb{N}$ (the first case is $p = 5$, $j = 14$). For these integer values of j, the characteristic number is zero for $l = 4p$ (thus, the first instance of $c_l^n = 0$ is for $n = 28, l = 20$).

In view of relation (6.49), we can define the *lower-middle* coherent state as

$$S_{-1/2}(\mathbf{n}) = S_{-1/2}^g \, , \text{ where } S_{-1/2} = \frac{1}{2} \left(\Pi_{n+2-[j+1/2]} + \Pi_{n+2-[j+1]} \right),$$

and the characteristic numbers of the symbol correspondence defined by this state, called the *lower-middle-state* symbol correspondence, satisfy

$$p_l^n(-1/2) = (-1)^l p_l^n(1/2). \tag{6.56}$$

Remark 6.2.57. Although the upper and lower-middle-state correspondences are mapping-positive symbol correspondences which stand in alternate relation to each other, we see from Proposition 6.2.54 that they are neither characteristic-positive nor characteristic-alternate correspondences, contrasting with the cases for the standard and alternate Berezin correspondences (highest and lowest states).

This same observation applies to the Berezin correspondences with characteristic numbers $b_l^n(2)$ and $b_l^n(n)$, cf. (6.48), which are always well defined when j is half-integer or when j is an integer different from $3, 6, 10, 15 \ldots (= l(l + 1)/2, l \geq 2)$.

However, apart from the fact that the characteristic numbers $p_l^n(1/2)$ never vanish, in contrast to $b_l^n(2)$, note from (6.48) that

$$\lim_{n \to \infty} b_l^n(2) = \lim_{n \to \infty} b_l^n = 1 \, ,$$

while

$$\lim_{n \to \infty} p_l^n(1/2) = (-1)^q \frac{(2q)!}{(q!)^2 2^{2q}} \, , \text{ when } l = 2q \text{ is even}$$

$$= 0 \, , \quad \text{when } l = 2q + 1 \text{ is odd} \tag{6.57}$$

and this other distinction shall be explored in Chap. 8.

Finally, we remark that there is a way to fully define the upper-middle coherent state in an *n*-invariant way, complementing Eq. (6.51), by recalling another standard definition from quantum mechanics:

Definition 6.2.58. If S is a state and $P \in \mathcal{B}(\mathcal{H}_j)$ is an operator, then

$$\sqrt{trace(SP^2) - (trace(SP))^2}$$

is the *standard deviation of the operator P in the state S*, or simply the *deviation of P in S*.

And then the following fact can be easily checked:

Proposition 6.2.59. *For any $n = 2j \in \mathbb{N}$, the upper-middle coherent state $S_{1/2}(\mathbf{n}) = S_{1/2}^g$, with $\mathbf{n} = g\mathbf{n}_0$ and $S_{1/2} = S_{1/2}(\mathbf{n}_0)$ given by (6.50), is the unique coherent state of minimal deviation for $J(\mathbf{n}) = J_3^g$ that satisfies (6.51).*

Namely, when j is half-integral, the deviation is zero, and when j is integral, the deviation is $1/2$. For any other coherent state $S'(\mathbf{n})$ satisfying (6.51), the deviation of $J(\mathbf{n})$ in that state will be higher, for any $n = 2j \in \mathbb{N}$.

Of course, the same definition applies for the lower-middle coherent state, with the obvious change in the value of (6.51) from $1/2$ to $-1/2$.

Chapter 7
Multiplications of Symbols on the 2-Sphere

Given any symbol correspondence $W^j = W^j_{\vec{c}}$, the algebra of operators in $\mathcal{B}(\mathcal{H}_j) \simeq M_{\mathbb{C}}(n+1)$ can be imported to the space of symbols $W^j_{\vec{c}}(\mathcal{B}(\mathcal{H}_j)) \simeq Poly_{\mathbb{C}}(S^2)_{\leq n} \subset C^\infty_{\mathbb{C}}(S^2)$. The 2-sphere, with such an algebra on a subset of its function space, has become known as the "fuzzy sphere" [47]. However, there is no single "fuzzy sphere", as each symbol correspondence defined by characteristic numbers $\vec{c} = (c_1, \ldots, c_n)$ gives rise to a distinct (although isomorphic) algebra on the space of symbols $Poly_{\mathbb{C}}(S^2)_{\leq n}$, as we shall investigate in some detail, in this chapter.

This fact has important bearings on the question of how these various symbol algebras, or $\vec{c}$-twisted j-algebras as they shall be called, relate to the classical Poisson algebra in the limit $n = 2j \to \infty$, as we shall see in Chap. 8.

7.1 Twisted Products of Spherical Symbols

Definition 7.1.1. For a given symbol correspondence (6.2) with symbol map $W^j = W^j_{\vec{c}}$, the *twisted product* $\star$ of symbols is the binary operation on symbols

$$\star : W^j_{\vec{c}}(\mathcal{B}(\mathcal{H}_j)) \times W^j_{\vec{c}}(\mathcal{B}(\mathcal{H}_j)) \to W^j_{\vec{c}}(\mathcal{B}(\mathcal{H}_j)) \simeq Poly_{\mathbb{C}}(S^2)_{\leq n}$$

induced by the product of operators via $W^j = W^j_{\vec{c}}$, that is, given by

$$W^j_{PQ} = W^j_P \star W^j_Q \,, \tag{7.1}$$

for every $P, Q \in \mathcal{B}(\mathcal{H}_j)$, where $W^j_P = W^j_{\vec{c}}(P)$, etc.

© Springer International Publishing Switzerland 2014
P. de M. Rios, E. Straume, *Symbol Correspondences for Spin Systems*,
DOI 10.1007/978-3-319-08198-4_7

Remark 7.1.2. We often write $\star^n$ and $\star_{\vec{c}}^n$ instead of $\star$ for the twisted product of symbols to emphasize its dependence on $n = 2j$ and $\vec{c} = (c_1, \ldots, c_n)$. Also, we follow [73] in calling the product induced by a symbol correspondence a *twisted product* of symbols and not a *star-product* of functions, or formal power series, because the latter product, introduced in [9], is defined from the classical data on S^2, as opposed to the former. The relationship between these two different kinds of product, in the limit $n \to \infty$, will be briefly commented later on, in the concluding chapter.

The following properties are immediate from Definitions 6.1.1 and 7.1.1:

Proposition 7.1.3. *The algebra of symbols defined by any twisted product is:*

$$
\begin{array}{lll}
(i) & \textit{An SO(3)-equivariant algebra}: (f_1 \star f_2)^g = f_1^g \star f_2^g, \\
(ii) & \textit{An associative algebra}: (f_1 \star f_2) \star f_3 = f_1 \star (f_2 \star f_3), \\
(iii) & \textit{A unital algebra}: 1 \star f = f \star 1 = f, \\
(iv) & \textit{A star algebra}: \overline{f_1 \star f_2} = \overline{f_2} \star \overline{f_1},
\end{array} \qquad (7.2)
$$

where $g \in SO(3)$ and $f_1, f_2, f_3 \in W^j(\mathcal{B}(\mathcal{H}_j)) \simeq Poly_{\mathbb{C}}(S^2)_{\leq n} \subset C_{\mathbb{C}}^{\infty}(S^2)$.

Remark 7.1.4. In view of Definition 7.1.1, we can use the normalization postulate (iv) in Definition 6.1.1 to define an induced inner product on the space of symbols:

$$
\left\langle W_P^j, W_Q^j \right\rangle_{\star} := \frac{1}{4\pi} \int_{S^2} \overline{W_P^j} \star W_Q^j \, dS. \qquad (7.3)
$$

With this induced inner product, the isometry postulate can be rewritten as:

$$
(v) \textit{ Isometry}: \left\langle W_P^j, W_Q^j \right\rangle = \left\langle W_P^j, W_Q^j \right\rangle_{\star}.
$$

7.1.1 Standard Twisted Products of Cartesian Symbols on S^2

Definition 7.1.5. The twisted product obtained via the standard Stratonovich-Weyl correspondence is called the *standard twisted product* of symbols and is denoted by $\star_1^n$, or simply $\star_1$.

Thus, for $f, g \in W_1^j(\mathcal{B}(\mathcal{H}_j)) = W^j(\mathcal{B}(\mathcal{H}_j)) \simeq Poly_{\mathbb{C}}(S^2)_{\leq n} \subset C_{\mathbb{C}}^{\infty}(S^2)$

$$
\star_1 = \star_1^n : (f, g) \mapsto f \star_1^n g = f \star_1 g \in W^j(\mathcal{B}(\mathcal{H}_j)) \simeq Poly_{\mathbb{C}}(S^2)_{\leq n} \subset C_{\mathbb{C}}^{\infty}(S^2).
$$

For a given symbol correspondence W, a basic problem is to calculate the twisted product of the cartesian coordinate functions x, y, z and exhibit its dependence on the spin parameter j. These coordinate functions form a basis for the linear symbols on S^2, i.e. homogeneous polynomials in x, y, z of degree 1.

Here we shall do this calculation in the basic case $W = W_1$, that is, the standard Stratonovich-Weyl correspondence. In the initial case $n = 2j = 1$, it follows from Example 3.3.8, the identities (4.30), and (6.11) with $\varepsilon_l = 1$,

$$x \longleftrightarrow \frac{1}{\sqrt{3}} \begin{pmatrix} 0 & 1 \\ 1 & 0 \end{pmatrix}, \, y \longleftrightarrow \frac{i}{\sqrt{3}} \begin{pmatrix} 0 & -1 \\ 1 & 0 \end{pmatrix}, \, z \longleftrightarrow \frac{1}{\sqrt{3}} \begin{pmatrix} 1 & 0 \\ 0 & -1 \end{pmatrix},$$

from which one deduces

$$x \star_1^1 y = -y \star_1^1 x = \frac{i}{\sqrt{3}} z, \quad x \star_1^1 x = \frac{1}{3}$$

and these identities hold after a cyclic permutation of (x, y, z). For $j \geq 1$ there is the following general result.

Proposition 7.1.6. *For the standard Stratonovich-Weyl correspondence, let* $n = 2j \geq 2$, *and for* $\{a, b, c\} = \{x, y, z\}$ *let* $\varepsilon_{abc} = \pm 1$ *be the sign of the permutation* $(x, y, z) \longrightarrow (a, b, c)$. *Then*

$$a \star_1^n b = \pi_n \cdot (ab) + \frac{\varepsilon_{abc} i}{\sqrt{n(n + 2)}} c, \tag{7.4}$$

$$a \star_1^n a = \pi_n \cdot (a^2) + \frac{1 - \pi_n}{3}, \tag{7.5}$$

$$a \star_1^n a + b \star_1^n b + c \star_1^n c = 1 \tag{7.6}$$

where

$$\pi_n = \frac{\sqrt{30}\mu_2}{\sqrt{n + 1}\, n(n + 2)} = \sqrt{\frac{(n - 1)(n + 3)}{n(n + 2)}}. \tag{7.7}$$

The proof of Proposition 7.1.6 follows readily from the general formula for the standard twisted product of spherical harmonics to be presented in the next section (cf. (7.10)), but in Appendix 10.7 the reader can find a more direct proof.

Remark 7.1.7. According to (10.41), via the standard Stratonovich-Weyl correspondence, the cartesian coordinate functions are identified (modulo a scaling) with the angular momentum operators in the same coordinate directions, namely

$$x = \frac{1}{\sqrt{j(j + 1)}} J_1, \, y = \frac{1}{\sqrt{j(j + 1)}} J_2, \, z = \frac{1}{\sqrt{j(j + 1)}} J_3. \tag{7.8}$$

On the other hand, for a symbol correspondence with characteristic numbers c_l^n one must also divide the operator on the right side by c_1^n (see below).

7.1.2 Twisted Products for General Symbol Correspondences

In order to study general twisted products of spherical symbols more systematically, we start with the twisted product of spherical harmonics. According to (6.12) and (3.99), for a symbol correspondence $W_{\tilde{c}}^j$ with characteristic numbers c_l^n, $l = 1, 2, \ldots n = 2j$, denoting the twisted product $\star^n$ by $\star_{\tilde{c}}^n$ we have

$$Y_{l_1}^{m_1} \star_{\tilde{c}}^n Y_{l_2}^{m_2} \longleftrightarrow \frac{\mu_0^2}{c_{l_1}^n c_{l_2}^n} e^j(l_1, m_1) e^j(l_2, m_2)$$

$$= (-1)^{2j+m} \frac{\mu_0^2}{c_{l_1}^n c_{l_2}^n} \sum_{l=0}^{2j} \begin{bmatrix} l_1 & l_2 & l \\ m_1 & m_2 & -m \end{bmatrix} [j] \, e^j(l, m)$$

$$\longleftrightarrow (-1)^{2j+m} \frac{\mu_0^2}{c_{l_1}^n c_{l_2}^n} \sum_{l=0}^{2j} \begin{bmatrix} l_1 & l_2 & l \\ m_1 & m_2 & -m \end{bmatrix} [j] \, \frac{c_l^n}{\mu_0} Y_l^m,$$

where $m = m_1 + m_2$ and

$$\begin{bmatrix} l_1 & l_2 & l \\ m_1 & m_2 & -m \end{bmatrix} [j]$$

is the Wigner product symbol given by Eqs. (3.98) and (3.104) in terms of Wigner $3jm$ and $6j$ symbols (cf. (3.105)–(3.109)). Thus, we arrive immediately at the following main result:

Theorem 7.1.8. *For $n = 2j \in \mathbb{N}$, let $W_{\tilde{c}}^j : \mathcal{B}(\mathcal{H}_j) \to Poly_{\mathbb{C}}(S^2)_{\leq n} \subset C_{\mathbb{C}}^\infty(S^2)$ be the symbol correspondence with characteristic numbers $c_l^n \neq 0$, that is,*

$$W_{\tilde{c}}^j : \mu_0 e^j(l, m) \longleftrightarrow c_l^n Y_l^m \, ; \quad -l \leq m \leq l \leq n \, .$$

Then, the corresponding twisted product of spherical harmonics $Y_{l_1}^{m_1}, Y_{l_2}^{m_2}$, induced by the operator product on $\mathcal{B}(\mathcal{H}_j)$, is given by

$$Y_{l_1}^{m_1} \star_{\tilde{c}}^n Y_{l_2}^{m_2} = (-1)^{n+m} \sqrt{n+1} \sum_{l=0}^{n} \begin{bmatrix} l_1 & l_2 & l \\ m_1 & m_2 & -m \end{bmatrix} [j] \, \frac{c_l^n}{c_{l_1}^n c_{l_2}^n} Y_l^m \qquad (7.9)$$

where $m = m_1 + m_2$ and $\delta(l_1, l_2, l) = 1$.

Corollary 7.1.9. *The standard twisted product of spherical harmonics is given by*

$$Y_{l_1}^{m_1} \star_1^n Y_{l_2}^{m_2} = (-1)^{n+m} \sqrt{n+1} \sum_{l=0}^{n} \begin{bmatrix} l_1 & l_2 & l \\ m_1 & m_2 & -m \end{bmatrix} [j] \, Y_l^m \, . \qquad (7.10)$$

and the twisted product induced by the standard Berezin correspondence defined by the characteristic numbers $b_l^n \in \mathbb{R}^+$ as in (6.37) is given by Eq. (7.9) above, via the following substitution:

$$
\frac{c_{l_3}^n}{c_{l_1}^n c_{l_2}^n} \rightarrow \frac{b_{l_3}^n}{b_{l_1}^n b_{l_2}^n} = \sqrt{\frac{\binom{n+l_1+1}{l_1}\binom{n+l_2+1}{l_2}\binom{n}{l_3}}{\binom{n}{l_1}\binom{n}{l_2}\binom{n+l_3+1}{l_3}}}
\tag{7.11}
$$

$$
= \sqrt{\frac{(n+l_1+1)!(n+l_2+1)!(n-l_1)!(n-l_2)!}{(n+l_3+1)!(n-l_3)!(n+1)!n!}} \,,
$$

Remark 7.1.10. (i) By linearity, given two symbols, $f = \sum_{l,m}\phi_{lm}Y_l^m$ and $g = \sum_{l,m}\gamma_{lm}Y_l^m$, their twisted product expands as

$$
f \star_c^n g = \sum_{l_1,l_2=0}^{n} \sum_{m_1=-l_1}^{l_1} \sum_{m_2=-l_2}^{l_2} \phi_{l_1 m_1}\gamma_{l_2 m_2} Y_{l_1}^{m_1} \star_c^n Y_{l_2}^{m_2} \,,
\tag{7.12}
$$

where $Y_{l_1}^{m_1} \star_c^n Y_{l_2}^{m_2}$ is given by formula (7.9), and also by the integral formula (7.101) below. Expanding the product (7.12) as $f \star_c^n g = \sum_{l,m}\rho_{lm}Y_l^m$, the coefficients $\rho_{lm} \in \mathbb{C}$ are given by

$$
(f \star_c^n g)_{lm} = \rho_{lm}
$$

$$
= (-1)^{n+m}\sqrt{n+1} \sum_{l_1,l_2=0}^{n} \sum_{m_1=-l_1}^{l_1} \begin{bmatrix} l_1 & l_2 & l \\ m_1 & m_2 & -m \end{bmatrix}[j] \frac{c_l^n}{c_{l_1}^n c_{l_2}^n} \phi_{l_1 m_1}\gamma_{l_2 m_2}
\tag{7.13}
$$

where $m_2 = m - m_1$ and the sum in l_1, l_2 is restricted by $\delta(l_1, l_2, l) = 1$.

(ii) By comparing (7.9) and (7.10), or directly from (7.13) above, we see that the expansion of any twisted product of spherical harmonics is immediately obtained from the expansion of the standard twisted product of spherical harmonics by multiplying each term of the standard expansion by $c_l^n/(c_{l_1}^n c_{l_2}^n)$.

To illustrate Theorem 7.1.8, Corollary 7.1.9 and Remark 7.1.10, let us first list the standard twisted product of linear spherical harmonics, which can be obtained directly from Theorem 7.1.8, or from Proposition 7.1.6 and Remark 7.1.10.

$$
Y_1^0 \star_1^n Y_1^0 = \pi_n \cdot \frac{2}{\sqrt{5}} Y_2^0 + 1,
\tag{7.14}
$$

$$
Y_1^{\pm 1} \star_1^n Y_1^{\pm 1} = \pi_n \cdot \sqrt{\frac{6}{5}} Y_2^{\pm 2},
\tag{7.15}
$$

$$Y_1^0 \star_1^n Y_1^{\pm 1} = \pi_n \cdot \sqrt{\frac{3}{5}} Y_2^{\pm 1} \pm \sqrt{\frac{3}{n(n+2)}} Y_1^{\pm 1}, \tag{7.16}$$

$$Y_1^{\pm 1} \star_1^n Y_1^0 = \pi_n \cdot \sqrt{\frac{3}{5}} Y_2^{\pm 1} \mp \sqrt{\frac{3}{n(n+2)}} Y_1^{\pm 1}, \tag{7.17}$$

$$Y_1^{\pm 1} \star_1^n Y_1^{\mp 1} = \pi_n \cdot \frac{1}{\sqrt{5}} Y_2^0 \mp \sqrt{\frac{3}{n(n+2)}} Y_1^0 - 1. \tag{7.18}$$

Then, from Remark 7.1.10, or directly from Corollary 7.1.9 we obtain that the standard Berezin twisted product of linear spherical harmonics is given by

$$Y_1^0 \star_b^n Y_1^0 = \frac{n-1}{n} \cdot \frac{2}{\sqrt{5}} Y_2^0 + \frac{n+2}{n}, \tag{7.19}$$

$$Y_1^{\pm 1} \star_b^n Y_1^{\pm 1} = \frac{n-1}{n} \cdot \sqrt{\frac{6}{5}} Y_2^{\pm 2}, \tag{7.20}$$

$$Y_1^0 \star_b^n Y_1^{\pm 1} = \frac{n-1}{n} \cdot \sqrt{\frac{3}{5}} Y_2^{\pm 1} \pm \frac{1}{n} \sqrt{3} Y_1^{\pm 1}, \tag{7.21}$$

$$Y_1^{\pm 1} \star_b^n Y_1^0 = \frac{n-1}{n} \cdot \sqrt{\frac{3}{5}} Y_2^{\pm 1} \mp \frac{1}{n} \sqrt{3} Y_1^{\pm 1}, \tag{7.22}$$

$$Y_1^{\pm 1} \star_b^n Y_1^{\mp 1} = \frac{n-1}{n} \cdot \frac{1}{\sqrt{5}} Y_2^0 \mp \frac{1}{n} \sqrt{3} Y_1^0 - \frac{n+2}{n}. \tag{7.23}$$

Thus, again from Remark 7.1.10, we finally get

Corollary 7.1.11. *The standard Berezin twisted product of the cartesian coordinate functions* $\{a, b, c\} = \{x, y, z\}$ *is given by*

$$a \star_b^n b = \frac{n-1}{n} ab + \frac{i \varepsilon_{abc}}{n} c, \tag{7.24}$$

$$a \star_b^n a = \frac{n-1}{n} a^2 + \frac{1}{n}, \tag{7.25}$$

$$a \star_b^n a + b \star_b^n b + c \star_b^n c = \frac{n+2}{n} = \frac{j+1}{j}. \tag{7.26}$$

Remark 7.1.12. Note, in particular, the distinction between Eq. (7.26) for the Berezin product and Eq. (7.6) for the standard Stratonovich-Weyl twisted product. In other words, from Remark 7.1.7 and (6.37), for the standard Berezin correspondence we have the identifications (compare with (7.8)):

$$x = \frac{1}{j+1} J_1, \quad y = \frac{1}{j+1} J_2, \quad z = \frac{1}{j+1} J_3.$$

The Parity Property for Symbols

The relationship between twisted products for the standard and the alternate Stratonovich-Weyl (resp. Berezin) symbol correspondences is described as follows:

Proposition 7.1.13. *Let W_1^j be the alternate Stratonovich-Weyl symbol correspondence given by characteristic numbers $c_l^n \equiv \varepsilon_l = (-1)^l$ (cf. Definition 6.2.38), and denote by $\star_{1-}^n$ its corresponding twisted product. Similarly, let B_-^j be the alternate Berezin symbol correspondence B_-^j, with characteristic numbers $b_{l-}^n = (-1)^l b_l^n$ (cf. Definition 6.2.37), and denote by $\star_{b-}^n$ its corresponding twisted product. More generally, denote by $\star_{c-}^n$ the twisted product induced by the symbol correspondence with characteristic numbers $c_{l-}^n = (-1)^l c_l^n$. Then, we have that*

$$Y_{l_1}^{m_1} \star_{c-}^n Y_{l_2}^{m_2} = Y_{l_2}^{m_2} \star_c^n Y_{l_1}^{m_1}. \tag{7.27}$$

Proof. Using the symmetry property (3.110) for the Wigner product symbols, the above identities follow from Eq. (7.9) and the fact that $c_{l-}^n/c_{l_1-}^n c_{l_2-}^n = (-1)^{l-l_1-l_2} c_l^n/c_{l_1}^n c_{l_2}^n = (-1)^{l_1+l_2+l} c_l^n/c_{l_1}^n c_{l_2}^n$. $\qquad\square$

Now, the parity property for operators (Proposition 3.3.12) has a neat version for any twisted product of symbols, to be stated in the proposition below. But first, let us make some definitions.

Definition 7.1.14. For any point $\mathbf{n} \in S^2$ let $-\mathbf{n}$ denote its antipodal point. We say that $f \in C_\mathbb{C}^\infty(S^2)$ is *even* if

$$f(\mathbf{n}) = f(-\mathbf{n}), \ \forall \mathbf{n} \in S^2$$

and $f \in C_\mathbb{C}^\infty(S^2)$ is *odd* if

$$f(\mathbf{n}) = -f(-\mathbf{n}), \ \forall \mathbf{n} \in S^2.$$

In particular, the null function 0 is the only function that is both even and odd.

For a given symbol correspondence and associated twisted product $f \star g$ of symbols, let us denote its symmetric product (or anti-commutator) by

$$[[\, f, g \,]]_\star = f \star g + g \star f$$

while its commutator is denoted in the usual way by

$$[\, f, g \,]_\star = f \star g - g \star f.$$

With these preparations, we state the following result.

Proposition 7.1.15 (The Parity Property for symbols). *With regard to the parity of symbols, the behavior of the above products is expressed as follows:*

$$[[\ even, even\ ]]_\star = even \qquad [\ odd, odd\ ]_\star = odd,$$

$$[[\ odd, odd\ ]]_\star = even \qquad [\ even, even\ ]_\star = odd, \tag{7.28}$$

$$[[\ even, odd\ ]]_\star = odd \qquad [\ even, odd\ ]_\star = even,$$

where, for example, $[[\ even, even\ ]]_\star = even$ means that the anti-commutator of two even symbols is an even symbol, and so on.

Proof. From Eq. (4.31), note that if the symbol f is even, then it can be written as a linear combination of even spherical harmonics,

$$f = \sum_{k,m} \alpha(k,m)\, Y_{2k}^m$$

and if the symbol g is odd, then it can be written as a linear combination of odd spherical harmonics,

$$g = \sum_{k,m} \beta(k,m)\, Y_{2k+1}^m.$$

Therefore, the relations (7.28) follow immediately from (7.9) and (3.110). □

Algebra Isomorphisms

The fact that the parity property for operators (Proposition 3.3.12) can be generally re-expressed for symbols (Proposition 7.1.15) follows essentially from the fact that, via any symbol correspondence $W_{\vec{c}}^j$, the linear space $Poly_{\mathbb{C}}(S^2)_{\leq n} \subset C_{\mathbb{C}}^\infty(S^2)$ with its twisted product $\star_{\vec{c}}^n$ is isomorphic to the matrix algebra $M_{\mathbb{C}}(n+1)$.

Definition 7.1.16. For each symbol correspondence $W_{\vec{c}}^j$, with characteristic numbers $\vec{c} = (c_1^n, c_2^n, \ldots, c_n^n)$, the space of symbols $W_{\vec{c}}^j(\mathcal{B}(\mathcal{H}_j)) = Poly_{\mathbb{C}}(S^2)_{\leq n} \subset C_{\mathbb{C}}^\infty(S^2)$ with its usual addition and the twisted product $\star_{\vec{c}}^n$ shall be called the *twisted j-algebra* associated to $W_{\vec{c}}^j$, or simply the *$\vec{c}$-twisted j-algebra*.

Proposition 7.1.17. *All twisted j-algebras are isomorphic to each other (same j) and no twisted j-algebra is isomorphic to any twisted j'-algebra, for $j \neq j'$.*

Proof. This follows immediately since every $\vec{c}$-twisted j-algebra is isomorphic to $M_{\mathbb{C}}(n+1)$, whereas the latter is not isomorphic to $M_{\mathbb{C}}(n'+1)$, for $n \neq n'$. □

Definition 7.1.18. If $(Poly_{\mathbb{C}}(S^2)_{\leq n}, \star_{\vec{c}}^n)$ and $(Poly_{\mathbb{C}}(S^2)_{\leq n}, \star_{\vec{c}'}^n)$ are two distinct twisted j-algebras, then the isomorphisms

$$U_{\vec{c},\vec{c}'}^j = W_{\vec{c}'}^j \circ (W_{\vec{c}}^j)^{-1} : (Poly_{\mathbb{C}}(S^2)_{\leq n}, \star_{\vec{c}}^n) \longrightarrow (Poly_{\mathbb{C}}(S^2)_{\leq n}, \star_{\vec{c}'}^n), \tag{7.29}$$

$$V^j_{\vec{c}',\vec{c}} = (W^j_{\vec{c}})^{-1} \circ W^j_{\vec{c}'} : M_{\mathbb{C}}(n+1) \to M_{\mathbb{C}}(n+1) \qquad (7.30)$$

and their inverses $U^j_{\vec{c}',\vec{c}}$ and $V^j_{\vec{c},\vec{c}'}$, respectively, define the *transition operators*.

The transition operator $U^j_{\vec{c},\vec{c}'}$ on $Poly_{\mathbb{C}}(S^2)_{\leq n}$ (resp. $V^j_{\vec{c}',\vec{c}}$ on $M_{\mathbb{C}}(n+1)$) has properties reflecting the properties listed in Definition 6.1.1, such as $SO(3)$-equivariance and preservation of real functions (resp. Hermitian matrices). There is the following commutative diagram relating the two types of transition operators

$$
\begin{array}{ccccc}
& V^j_{\vec{c}',\vec{c}} & & Id & \\
M_{\mathbb{C}}(n+1) & \longrightarrow & M_{\mathbb{C}}(n+1) & \longrightarrow & M_{\mathbb{C}}(n+1) \\
W^j_{\vec{c}'} \downarrow & & \downarrow W^j_{\vec{c}} & & \downarrow W^j_{\vec{c}'} \\
& Id & & U^j_{\vec{c},\vec{c}'} & \\
Poly_{\mathbb{C}}(S^2)_{\leq n} & \longrightarrow & Poly_{\mathbb{C}}(S^2)_{\leq n} & \longrightarrow & Poly_{\mathbb{C}}(S^2)_{\leq n},
\end{array}
\qquad (7.31)
$$

Definition 7.1.19. The (small) category of general spin-j symbol correspondences, whose (invertible) morphisms are induced by the transition operators as in (7.31), shall be called the *dequantization groupoid* of a spin-j system (cf. Remark 6.2.1).

Remark 7.1.20. Although all twisted j-algebras are isomorphic (for fixed j), the fact that distinct symbol correspondences define distinct twisted j-algebras has nontrivial consequences when we consider sequences (in $n = 2j \in \mathbb{N}$) of twisted j-algebras and their limits as $n \to \infty$, as we shall see in the next chapter.

7.2 Integral Representations of Twisted Products

Just as the Moyal product of symbols on $\mathbb{R}^{2n}$ has an integral version, the Groenewold-von Neumann product [35, 77], it is interesting to see that twisted products of spherical symbols can also be written in integral form:

$$f \star g\,(\mathbf{n}) = \iint_{S^2 \times S^2} f(\mathbf{n}_1) g(\mathbf{n}_2) \mathbb{L}(\mathbf{n}_1, \mathbf{n}_2, \mathbf{n}) d\mathbf{n}_1 d\mathbf{n}_2 . \qquad (7.32)$$

Integral forms for twisted products in principle allow for a direct definition of general twisted products of arbitrary symbols $f, g \in Poly_{\mathbb{C}}(S^2)_{\leq n}$ without the need to decompose them in the basis of spherical harmonics.

In such an integral representation, all properties of the twisted product are encoded in the integral trikernel

$$\mathbb{L} : S^2 \times S^2 \times S^2 \to \mathbb{C}. \qquad (7.33)$$

Therefore, for each symbol correspondence $W_{\check{c}}^{j}$, there will be a distinct integral trikernel $\mathbb{L}_{\check{c}}^{j}$.

7.2.1 General Formulae and Properties of Integral Trikernels

For the standard Stratonovich-Weyl symbol correspondence W_1^j, from Eqs. (6.13), (6.24), and Proposition 6.2.20, we immediately have the following result:

Proposition 7.2.1. *Let $f, g \in C_{\mathbb{C}}^{\infty}(S^2)$ be such that $f = W_1^j(F)$, $g = W_1^j(G)$, for $F, G \in \mathcal{B}(\mathcal{H}_j)$, where W_1^j is determined by the operator kernel K_1^j with all characteristic numbers $c_l = 1$, for $0 \leq l \leq n = 2j$, in Eq. (6.16). Then,*

$$f \star_1^n g\,(\mathbf{n}) = \iint_{S^2 \times S^2} f(\mathbf{n}_1)g(\mathbf{n}_2)\mathbb{L}_1^j(\mathbf{n}_1, \mathbf{n}_2, \mathbf{n})d\,\mathbf{n}_1 d\,\mathbf{n}_2\,, \qquad (7.34)$$

where

$$\mathbb{L}_1^j(\mathbf{n}_1, \mathbf{n}_2, \mathbf{n}) = \left(\frac{2j+1}{4\pi}\right)^2 trace(K_1^j(\mathbf{n}_1)K_1^j(\mathbf{n}_2)K_1^j(\mathbf{n})) \qquad (7.35)$$

is the standard Stratonovich trikernel.

Corollary 7.2.2. *The Stratonovich trikernel is, in fact, a polynomial function $\mathbb{L}_1^j \in (Poly_{\mathbb{C}}(S^2)_{\leq n})^3 \subset C_{\mathbb{C}}^{\infty}(S^2 \times S^2 \times S^2)$ which can be written as*

$$\mathbb{L}_1^j(\mathbf{n}_1, \mathbf{n}_2, \mathbf{n}_3) \qquad (7.36)$$

$$= \frac{(-1)^n \sqrt{n+1}}{(4\pi)^2} \sum_{l_i, m_i} \begin{bmatrix} l_1 & l_2 & l_3 \\ m_1 & m_2 & m_3 \end{bmatrix}[j]\, \overline{Y_{l_1}^{m_1}}(\mathbf{n}_1)\overline{Y_{l_2}^{m_2}}(\mathbf{n}_2)\overline{Y_{l_3}^{m_3}}(\mathbf{n}_3)$$

with the summations in l_i and m_i subject to the constraints

$$0 \leq l_i \leq n = 2j,\ -l_i \leq m_i \leq l_i,\ \delta(l_1, l_2, l_3) = 1,\ m_1 + m_2 + m_3 = 0. \qquad (7.37)$$

Proof. This follows straightforwardly from the decomposition of symbols in the standard (orthonormal) basis of spherical harmonics, using formula (7.10) for the standard twisted product of basis vectors, and Eq. (4.28).   □

From the above formula, using (3.110) we immediately obtain:

Corollary 7.2.3. *The Stratonovich trikernel is symmetric, in the sense that it satisfies*

$$\mathbb{L}_1^j(\mathbf{n}_1, \mathbf{n}_2, \mathbf{n}_3) = \mathbb{L}_1^j(\mathbf{n}_3, \mathbf{n}_1, \mathbf{n}_2) = \mathbb{L}_1^j(\mathbf{n}_2, \mathbf{n}_3, \mathbf{n}_1)\,. \qquad (7.38)$$

General Formulae for Trikernels

For a general correspondence $W_{\vec{c}}^{j}$, with operator kernel $K_{\vec{c}}^{j}$ as in (6.16) and characteristic numbers $\vec{c} = (c_1^n, c_2^n, \ldots, c_n^n)$, we obtain from Eq. (6.13) and Theorem 6.2.21 the generalization of Proposition 7.2.1 and Corollary 7.2.2.

Theorem 7.2.4. *A general twisted product of $f, g \in Poly_{\mathbb{C}}(S^2)_{\leq n}$ is given by*

$$f \star_{\vec{c}}^{n} g \, (\mathbf{n}) = \iint_{S^2 \times S^2} f(\mathbf{n}_1) g(\mathbf{n}_2) \mathbb{L}_{\vec{c}}^{j}(\mathbf{n}_1, \mathbf{n}_2, \mathbf{n}) d\mathbf{n}_1 d\mathbf{n}_2 \, , \qquad (7.39)$$

with

$$\mathbb{L}_{\vec{c}}^{j}(\mathbf{n}_1, \mathbf{n}_2, \mathbf{n}_3) = \left(\frac{2j+1}{4\pi} \right)^2 trace(\tilde{K}_{\vec{c}}^{j}(\mathbf{n}_1) \tilde{K}_{\vec{c}}^{j}(\mathbf{n}_2) K_{\vec{c}}^{j}(\mathbf{n}_3)) \qquad (7.40)$$

$$= \frac{(-1)^n \sqrt{n+1}}{(4\pi)^2} \sum_{l_i, m_i} \begin{bmatrix} l_1 & l_2 & l_3 \\ m_1 & m_2 & m_3 \end{bmatrix} [j] \frac{c_{l_3}^n}{c_{l_1}^n c_{l_2}^n} \overline{Y_{l_1}^{m_1}}(\mathbf{n}_1) \overline{Y_{l_2}^{m_2}}(\mathbf{n}_2) \overline{Y_{l_3}^{m_3}}(\mathbf{n}_3)$$

as the corresponding integral trikernel, where $K_{\vec{c}}^{j}$ and $\tilde{K}_{\vec{c}}^{j}$ are the operator kernels given by Eq. (6.16) with characteristic numbers $\vec{c} = (c_l^n) = (c_1^n, \ldots, c_n^n)$ and $\frac{1}{\vec{c}} = (\frac{1}{c_1^n}, \ldots, \frac{1}{c_n^n})$, respectively, and with the restrictions (7.37) for the l_i, m_i summations.

Note that the trikernel $\mathbb{L}_{\vec{c}}^{j}$ is a polynomial function, and in general it is not symmetric, namely (7.38) does not hold. More explicitly, the trikernel (7.40) has the full expression

$$\mathbb{L}_{\vec{c}}^{j}(\mathbf{n}_1, \mathbf{n}_2, \mathbf{n}_3) \qquad (7.41)$$

$$= \frac{\sqrt{n+1}}{(4\pi)^2} \sum_{l_i} \sqrt{\frac{(n-l_1)!(n-l_2)!(n-l_3)!}{(n+l_1+1)!(n+l_2+1)!(n+l_3+1)!}} \Delta^2(l_1, l_2, l_3) l_1! l_2! l_3!$$

$$\cdot \frac{c_{l_3}^n}{c_{l_1}^n c_{l_2}^n} \sqrt{(2l_1+1)(2l_2+1)(2l_3+1)} \sum_{k} \frac{(-1)^k (n+1+k)!}{(n+k-l_1-l_2-l_3)! R(l_1, l_2, l_3; k)}$$

$$\cdot \sum_{m_i} S_{m_1, m_2, m_3}^{l_1, \, l_2, \, l_3} N_{m_1, m_2, m_3}^{l_1, l_2, l_3} Y_{l_1}^{m_1}(\mathbf{n}_1) Y_{l_2}^{m_2}(\mathbf{n}_2) Y_{l_3}^{m_3}(\mathbf{n}_3)$$

with $\Delta(l_1, l_2, l_3), S_{m_1, m_2, m_3}^{l_1, \, l_2, \, l_3}, N_{m_1, m_2, m_3}^{l_1, l_2, l_3}$, and $R(l_1, l_2, l_3; k)$ given respectively by (3.48), (3.49), (3.106), and (3.109), with restrictions on the summations over l_i, m_i, k according to (7.37) and Remark 3.2.8.

In particular, the explicit expression for the Stratonovich trikernel is obtained from (7.41) by setting all $c_l^n = 1$ and we obtain explicit expressions for the integral trikernel $\mathbb{L}_{\vec{b}}^{j}$ of the standard Berezin twisted product by performing the substitution (7.11) in (7.40)–(7.41).

General Properties of Trikernels

Since $\mathbb{L}_{\overset{j}{c}}$ is a polynomial function it also follows from the general formulae (7.39)–(7.41) that we cannot use the integral formulation to extend a twisted product defined on $Poly_{\mathbb{C}}(S^2)_{\leq n}$ to a larger subset of $C_{\mathbb{C}}^{\infty}(S^2)$, namely we have:

Corollary 7.2.5. *Let $f, g \in C_{\mathbb{C}}^{\infty}(S^2)$, $\mathcal{L} \in C_{\mathbb{C}}^{\infty}(S^2 \times S^2 \times S^2)$, and define a binary operation $\bullet : C_{\mathbb{C}}^{\infty}(S^2) \times C_{\mathbb{C}}^{\infty}(S^2) \to C_{\mathbb{C}}^{\infty}(S^2)$ via the integral formula*

$$f \bullet g \, (\mathbf{n}) = \int_{S^2 \times S^2} f(\mathbf{n}_1) g(\mathbf{n}_2) \mathcal{L}(\mathbf{n}_1, \mathbf{n}_2, \mathbf{n}) d\mathbf{n}_1 d\mathbf{n}_2 \,. \tag{7.42}$$

If $\mathcal{L} = \mathbb{L}_{\overset{j}{c}} \in (Poly_{\mathbb{C}}(S^2)_{\leq n})^3 \subset C_{\mathbb{C}}^{\infty}(S^2 \times S^2 \times S^2)$, then $\bullet = \star_{\overset{n}{c}} : Poly_{\mathbb{C}}(S^2)_{\leq n} \times Poly_{\mathbb{C}}(S^2)_{\leq n} \to Poly_{\mathbb{C}}(S^2)_{\leq n}$ and, in particular, if either f or $g \in C_{\mathbb{C}}^{\infty}(S^2) \setminus Poly_{\mathbb{C}}(S^2)_{\leq n}$, then $f \bullet g = 0$.

Proof. We prove that, if $\mathcal{L} = \mathbb{L}_{\overset{j}{c}} \in (Poly_{\mathbb{C}}(S^2)_{\leq n})^3$, then $f \bullet g = 0$ if either f or $g \in C_{\mathbb{C}}^{\infty}(S^2) \setminus Poly_{\mathbb{C}}(S^2)_{\leq n}$. From this it follows immediately that $\bullet = \star_{\overset{n}{c}}$.

Now, if $g \in C_{\mathbb{C}}^{\infty}(S^2) \setminus Poly_{\mathbb{C}}(S^2)_{\leq n}$, then g can be expanded as a series of Y_l^m's, with all $l > n$. But from (7.39)–(7.40), with the restriction (7.37), $f \bullet g = 0$ follows from orthogonality of each Y_l^m in the g series and every $Y_{l'}^{m'}$, for $n < l \neq l' \leq n$. $\qquad\square$

The integral trikernels $\mathbb{L}$ of twisted products have some common properties that we shall spell out, as follows.

Proposition 7.2.6. *If $\mathbb{L} = \mathbb{L}_{\overset{j}{c}}$ is the integral trikernel of a twisted product according to Eq. (7.32), then it satisfies:*

(i) $\mathbb{L}(\mathbf{n}_1, \mathbf{n}_2, \mathbf{n}) = \mathbb{L}(g\mathbf{n}_1, g\mathbf{n}_2, g\mathbf{n})$,

(ii) $\displaystyle\int_{S^2} \mathbb{L}(\mathbf{n}_1, \mathbf{n}_2, \mathbf{n}) \mathbb{L}(\mathbf{n}, \mathbf{n}_3, \mathbf{n}_4) d\mathbf{n} = \int_{S^2} \mathbb{L}(\mathbf{n}_1, \mathbf{n}, \mathbf{n}_4) \mathbb{L}(\mathbf{n}_2, \mathbf{n}_3, \mathbf{n}) d\mathbf{n}$,

(iii) $\displaystyle\int_{S^2} \mathbb{L}(\mathbf{n}_1, \mathbf{n}_2, \mathbf{n}) d\mathbf{n}_1 = R^j(\mathbf{n}_2, \mathbf{n}), \; \int_{S^2} \mathbb{L}(\mathbf{n}_1, \mathbf{n}_2, \mathbf{n}) d\mathbf{n}_2 = R^j(\mathbf{n}_1, \mathbf{n})$,

(iv) $\mathbb{L}(\mathbf{n}_2, \mathbf{n}_1, \mathbf{n}) = \overline{\mathbb{L}(\mathbf{n}_1, \mathbf{n}_2, \mathbf{n})}$,

where $g \in SO(3)$, and $R^j(\mathbf{n}_2, \mathbf{n}) \in (Poly_{\mathbb{C}}(S^2)_{\leq n})^2$ is the reproducing kernel of the truncated polynomial algebra (twisted j-algebra) $Poly_{\mathbb{C}}(S^2)_{\leq n}$, characterized by

$$\int_{S^2} R^j(\mathbf{n}_2, \mathbf{n}) f(\mathbf{n}_2) d\mathbf{n}_2 = f(\mathbf{n}) \,, \; \forall f \in Poly_{\mathbb{C}}(S^2)_{\leq n}. \tag{7.43}$$

Corollary 7.2.7. *The reproducing kernel has the expansion*

$$R^j(\mathbf{n}_1, \mathbf{n}_2) = \frac{1}{4\pi} \sum_{l=0}^{2j} \sum_{m=-l}^{l} \overline{Y_l^m}(\mathbf{n}_1) Y_l^m(\mathbf{n}_2) \tag{7.44}$$

$$= \frac{1}{4\pi} \sum_{l=0}^{2j} (2l+1) P_l(\mathbf{n}_1 \cdot \mathbf{n}_2) = R^j(\mathbf{n}_2, \mathbf{n}_1) \tag{7.45}$$

where functions P_l are the Legendre polynomials (see Sect. 4.2.2, also [74]), and $\mathbf{n}_1 \cdot \mathbf{n}_2$ denotes the euclidean inner product of unit vectors in $\mathbb{R}^3$. In particular, $R^j(\mathbf{n}_1, \mathbf{n}_2) = R^j(\mathbf{n}_2, \mathbf{n}_1)$.

Proof. From Eq. (7.32), properties (i)–(iv) in Proposition 7.2.6 follow directly from properties (i)–(iv) in Proposition 7.1.3, together with Corollary 7.2.5.

Next, we observe that (7.44) is equivalent to (7.43). But (7.44) follows from Eq. (7.40) and property (iii) in Proposition 7.2.6, using that

$$\int_{S^2} Y_m^l(\mathbf{n}) d\mathbf{n} = 0, \text{ if } (l,m) \neq (0,0),$$

together with the identity

$$\begin{pmatrix} 0 & l & l \\ 0 & -m & m \end{pmatrix} \begin{Bmatrix} 0 & l & l \\ j & j & j \end{Bmatrix} = \frac{(-1)^{2j+m}}{\sqrt{(2j+1)(2l+1)}}$$

(cf. Eqs. (3.94) and (3.104), and equations 8.5.1 and 9.5.1 in [74]).

Finally, to obtain (7.45) we use the identity

$$(2l+1) P_l(\mathbf{n}_1 \cdot \mathbf{n}_2) = \sum_{m=-l}^{l} \overline{Y_l^m}(\mathbf{n}_1) Y_l^m(\mathbf{n}_2), \tag{7.46}$$

which follows from Eqs. (4.32) and (4.47). ☐

Remark 7.2.8. (i) If $\mathbb{L}_{\frac{j}{c}}^j$ is the integral trikernel of a characteristic-positive symbol correspondence, then by (7.27), (7.39), and Proposition 7.2.6, property (iv), the integral trikernel of the characteristic-alternate symbol correspondence is given by

$$\mathbb{L}_{\frac{j}{c-}} = \overline{\mathbb{L}_{\frac{j}{c}}}. \tag{7.47}$$

(ii) For a symmetric integral trikernel $\mathbb{L}^j$ (cf. Eq. (7.38)), as for example the standard Stratonovich trikernel (cf. Eq. (7.36)), we also have that

$$\int_{S^2} \mathbb{L}_1^j(\mathbf{n}_1, \mathbf{n}_2, \mathbf{n}) d\mathbf{n} = R^j(\mathbf{n}_1, \mathbf{n}_2), \qquad (7.48)$$

but this identity does not hold for nonsymmetric integral trikernels.

Explicitly $SO(3)$-Invariant Formulae for Trikernels

Now, we turn to the $SO(3)$-invariance of the integral trikernel, namely, according to property (i) in Proposition 7.2.6, $\mathbb{L}_{\overline{c}}^j$ can also be expressed in terms of $SO(3)$-invariant functions. First of all, let us recall the following basic fact, whose proof is found in a more general setting in Weyl [87].

Lemma 7.2.9. *Every $SO(3)$-invariant function of three points on the 2-sphere S^2 represented by unit vectors $\mathbf{n}_i, i = 1, 2, 3$, in euclidean 3-space, can be expressed as a function of the three euclidean inner products $\mathbf{n}_1 \cdot \mathbf{n}_2, \mathbf{n}_2 \cdot \mathbf{n}_3, \mathbf{n}_3 \cdot \mathbf{n}_1$, together with the determinant*

$$[\mathbf{n}_1, \mathbf{n}_2, \mathbf{n}_3] = \det(\mathbf{n}_1, \mathbf{n}_2, \mathbf{n}_3) . \qquad (7.49)$$

Thus, we introduce two $SO(3)$-invariant functions of type (7.33), namely

$$\mathcal{T}(\mathbf{n}_1, \mathbf{n}_2, \mathbf{n}_3) = (\mathbf{n}_1 \cdot \mathbf{n}_2) - (\mathbf{n}_1 \cdot \mathbf{n}_3)(\mathbf{n}_2 \cdot \mathbf{n}_3) - i[\mathbf{n}_1, \mathbf{n}_2, \mathbf{n}_3] , \quad i = \sqrt{-1} ,$$

and the following function depending on three integers $l_k \geq 0$:

$$\mathcal{L}_{l_1, l_2, l_3}(\mathbf{n}_1, \mathbf{n}_2, \mathbf{n}_3) \qquad (7.50)$$

$$= (2l_1 + 1)(2l_2 + 1)(2l_3 + 1) \cdot \left[\begin{pmatrix} l_1 & l_2 & l_3 \\ 0 & 0 & 0 \end{pmatrix} P_{l_1}(\mathbf{n}_1 \cdot \mathbf{n}_3) P_{l_2}(\mathbf{n}_2 \cdot \mathbf{n}_3) \right.$$

$$+ \sum_{m=1}^{\min\{l_1, l_2\}} (-1)^m \begin{pmatrix} l_1 & l_2 & l_3 \\ m & -m & 0 \end{pmatrix} \prod_{k=1}^{2} \sqrt{\frac{(l_k - m)!}{(l_k + m)!}} \frac{P_{l_k}^m(\mathbf{n}_k \cdot \mathbf{n}_3)}{(1 - (\mathbf{n}_k \cdot \mathbf{n}_3)^2)^{m/2}}$$

$$\left. \cdot \{(\mathcal{T}(\mathbf{n}_1, \mathbf{n}_2, \mathbf{n}_3))^m + (-1)^L (\mathcal{T}(\mathbf{n}_2, \mathbf{n}_1, \mathbf{n}_3))^m\} \right]$$

where $L = l_1 + l_2 + l_3$. In particular, $\mathcal{L}_{l_1, l_2, l_3}$ has the following property which is obvious from (7.50) and the fact that

$$L \text{ is odd} \Rightarrow \begin{pmatrix} l_1 & l_2 & l_3 \\ 0 & 0 & 0 \end{pmatrix} = 0 .$$

Lemma 7.2.10. *$\mathcal{L}_{l_1, l_2, l_3}$ is real when L is even and $\mathcal{L}_{l_1, l_2, l_3}$ is purely imaginary when L is odd.*

With these preparations, we can state the following result.

Theorem 7.2.11. *The $\vec{c}$-correspondence trikernel $\mathbb{L}_{\vec{c}}^{j}$ can also be written as*

$$\mathbb{L}_{\vec{c}}^{j}(\mathbf{n}_1, \mathbf{n}_2, \mathbf{n}_3) \tag{7.51}$$

$$= (-1)^{2j} \frac{\sqrt{2j+1}}{(4\pi)^2} \sum_{\substack{l_1,l_2,l_3=0 \\ \delta(l_1,l_2,l_3)=1}}^{2j} \begin{Bmatrix} l_1 & l_2 & l_3 \\ j & j & j \end{Bmatrix} \frac{c_{l_3}^{n}}{c_{l_1}^{n} c_{l_2}^{n}} \mathcal{L}_{l_1,l_2,l_3}(\mathbf{n}_1, \mathbf{n}_2, \mathbf{n}_3).$$

Proof. By $SO(3)$-invariance, we may assume $\mathbf{n}_3 = \mathbf{n}_0 = (0, 0, 1) \in \mathbb{R}^3$. Then the formula (7.51), including the expression (7.50) for $\mathcal{L}_{l_1,l_2,l_3}$, are obtained in a rather straightforward way from formulae (7.40)–(7.41).   □

Remark 7.2.12. Formula (7.50) for the function $\mathcal{L}_{l_1,l_2,l_3}$ looks asymmetric with respect to l_3 and $\mathbf{n}_3$, in comparison with l_1, l_2 and $\mathbf{n}_1, \mathbf{n}_2$. However, such asymmetry is only apparent and reflects the asymmetrical way in which the formula for $\mathcal{L}$ was derived. The situation resembles some well-known formulae for the Wigner $6j$ symbol, which do not manifestly exhibit some of the symmetries of the symbol. Indeed, we have the following symmetry properties when (l_1, l_2, l_3) and $(\mathbf{n}_1, \mathbf{n}_2, \mathbf{n}_3)$ are permuted covariantly:

$$\mathcal{L}_{l_1,l_2,l_3}(\mathbf{n}_1, \mathbf{n}_2, \mathbf{n}_3) = (-1)^L \mathcal{L}_{l_2,l_1,l_3}(\mathbf{n}_2, \mathbf{n}_1, \mathbf{n}_3) = (-1)^L \mathcal{L}_{l_1,l_3,l_2}(\mathbf{n}_1, \mathbf{n}_3, \mathbf{n}_2). \tag{7.52}$$

Example 7.2.13. In view of Eq. (7.52) above, we list the functions $\mathcal{L}_{l_1,l_2,l_3}$ for $l_k \le 2$ subject to $\delta(l_1, l_2, l_3) = 1$, as follows:

$$\mathcal{L}_{0,0,0}(\mathbf{n}_1, \mathbf{n}_2, \mathbf{n}_3) = 1,$$

$$\mathcal{L}_{1,1,0}(\mathbf{n}_1, \mathbf{n}_2, \mathbf{n}_3) = -3\sqrt{3}(\mathbf{n}_1 \cdot \mathbf{n}_2),$$

$$\mathcal{L}_{1,1,1}(\mathbf{n}_1, \mathbf{n}_2, \mathbf{n}_3) = i9\sqrt{\frac{3}{2}}[\mathbf{n}_1, \mathbf{n}_2, \mathbf{n}_3],$$

$$\mathcal{L}_{2,1,1}(\mathbf{n}_1, \mathbf{n}_2, \mathbf{n}_3) = 3\sqrt{\frac{15}{2}} \{3(\mathbf{n}_1 \cdot \mathbf{n}_2)(\mathbf{n}_1 \cdot \mathbf{n}_3) - (\mathbf{n}_2 \cdot \mathbf{n}_3)\},$$

$$\mathcal{L}_{2,2,0}(\mathbf{n}_1, \mathbf{n}_2, \mathbf{n}_3) = 5\sqrt{5}P_2(\mathbf{n}_1 \cdot \mathbf{n}_2) = \frac{5\sqrt{5}}{2}\{3(\mathbf{n}_1 \cdot \mathbf{n}_2)^2 - 1\},$$

$$\mathcal{L}_{2,2,1}(\mathbf{n}_1, \mathbf{n}_2, \mathbf{n}_3) = -i15\sqrt{\frac{15}{2}}(\mathbf{n}_1 \cdot \mathbf{n}_2)[\mathbf{n}_1, \mathbf{n}_2, \mathbf{n}_3],$$

$$\mathcal{L}_{2,2,2}(\mathbf{n}_1, \mathbf{n}_2, \mathbf{n}_3) = -25\sqrt{\frac{5}{14}}\{3\{[\mathbf{n}_1, \mathbf{n}_2, \mathbf{n}_3]^2 + (\mathbf{n}_1 \cdot \mathbf{n}_2)(\mathbf{n}_2 \cdot \mathbf{n}_3)(\mathbf{n}_3 \cdot \mathbf{n}_1)\} - 1\}.$$

The following $SO(3)$-invariant function will be used in various formulas below, in this chapter, so we denote it here as

$$X(\mathbf{n}_1, \mathbf{n}_2, \mathbf{n}_3) = \mathbf{n}_1 \cdot \mathbf{n}_2 + \mathbf{n}_2 \cdot \mathbf{n}_3 + \mathbf{n}_3 \cdot \mathbf{n}_1. \tag{7.53}$$

Example 7.2.14. We can now calculate the Stratonovich trikernel $\mathbb{L}_1^j$ for the two values $j = 1/2, 1$, by substituting the appropriate expressions for $\mathcal{L}_{l_1, l_2, l_3}$ listed in Example 7.2.13 into Eq. (7.51):

$$\mathbb{L}_1^{1/2}(\mathbf{n}_1, \mathbf{n}_2, \mathbf{n}_3) = \frac{1}{(4\pi)^2}\{1 + 3X(\mathbf{n}_1, \mathbf{n}_2, \mathbf{n}_3) + i3\sqrt{3}[\mathbf{n}_1, \mathbf{n}_2, \mathbf{n}_3]\}, \tag{7.54}$$

$$\mathbb{L}_1^1(\mathbf{n}_1, \mathbf{n}_2, \mathbf{n}_3) = \frac{1}{(4\pi)^2}\left\{1 + 3X(\mathbf{n}_1, \mathbf{n}_2, \mathbf{n}_3) + \frac{15}{2}Z(\mathbf{n}_1, \mathbf{n}_2, \mathbf{n}_3)\right. \tag{7.55}$$

$$+ \frac{\sqrt{10}}{8}\left((3X(\mathbf{n}_1, \mathbf{n}_2, \mathbf{n}_3) - 1)^2 - 24Z(\mathbf{n}_1, \mathbf{n}_2, \mathbf{n}_3) - 45[\mathbf{n}_1, \mathbf{n}_2, \mathbf{n}_3]^2\right)$$

$$\left. + i\frac{9\sqrt{2}}{4}(1 + 5X(\mathbf{n}_1, \mathbf{n}_2, \mathbf{n}_3))[\mathbf{n}_1, \mathbf{n}_2, \mathbf{n}_3]\right\},$$

where $X(\mathbf{n}_1, \mathbf{n}_2, \mathbf{n}_3)$ is given by (7.53) and

$$Z(\mathbf{n}_1, \mathbf{n}_2, \mathbf{n}_3) = (\mathbf{n}_1 \cdot \mathbf{n}_2)^2 + (\mathbf{n}_2 \cdot \mathbf{n}_3)^2 + (\mathbf{n}_3 \cdot \mathbf{n}_1)^2 - 1.$$

Using Eq. (6.37) for the Berezin characteristic numbers $c_l^n = b_l^n$, we can similarly calculate the Berezin trikernel $\mathbb{L}_{\tilde{b}}^j$ for the two lowest values of j:

$$\mathbb{L}_{\tilde{b}}^{1/2}(\mathbf{n}_1, \mathbf{n}_2, \mathbf{n}_3) = \frac{1}{(4\pi)^2}\{1 + 3X^{1/2}(\mathbf{n}_1, \mathbf{n}_2, \mathbf{n}_3) + i9[\mathbf{n}_1, \mathbf{n}_2, \mathbf{n}_3]\} \tag{7.56}$$

$$\mathbb{L}_{\tilde{b}}^1(\mathbf{n}_1, \mathbf{n}_2, \mathbf{n}_3) = \frac{1}{(4\pi)^2}\left(\frac{9}{4}\right)\left\{(1 - X^1(\mathbf{n}_1, \mathbf{n}_2, \mathbf{n}_3))^2\right. \tag{7.57}$$

$$- 6((\mathbf{n}_2 \cdot \mathbf{n}_3)^2 + (\mathbf{n}_3 \cdot \mathbf{n}_1)^2 - 2(\mathbf{n}_1 \cdot \mathbf{n}_2)) - 25[\mathbf{n}_1, \mathbf{n}_2, \mathbf{n}_3]^2$$

$$\left. + i2(1 + 5X^1(\mathbf{n}_1, \mathbf{n}_2, \mathbf{n}_3))[\mathbf{n}_1, \mathbf{n}_2, \mathbf{n}_3]\right\},$$

where

$$X^j(\mathbf{n}_1, \mathbf{n}_2, \mathbf{n}_3) = (4j + 1)(\mathbf{n}_1 \cdot \mathbf{n}_2) + (\mathbf{n}_2 \cdot \mathbf{n}_3) + (\mathbf{n}_3 \cdot \mathbf{n}_1).$$

Except for $\mathbb{L}_1^{1/2}$, which was first presented in [67], and $\mathbb{L}_{\tilde{b}}^{1/2}$, which can be inferred from [92], the formulae in Examples 7.2.13 and 7.2.14 were originally obtained by Nazira Harb, who also worked out formulae for $\mathcal{L}$ for some higher values of l_i and formulae for $\mathbb{L}_1^j$ and $\mathbb{L}_{\tilde{b}}^j$ for some higher values of j.

7.2.2   *Recursive Trikernels and Transition Kernels*

Recursive Trikernels

A recursive trikernel associated to the standard Berezin correspondence has been defined by Wildberger [92]. This trikernel is symmetric, in the sense of Eq. (7.38), but is not the trikernel of a bona-fide integral equation for the twisted product in the sense of Eq. (7.39); rather, it is the trikernel of a recursive integral equation for the standard Berezin twisted product.

Let B be the standard Berezin symbol correspondence, given by Eq. (6.33) in Theorem 6.2.27. We recall from Remark 7.1.4 that, because this symbol correspondence is not isometric, there are two natural $SU(2)$-invariant inner products on the space of symbols $Poly_{\mathbb{C}}(S^2)_{\leq n}$, namely

- The usual L^2-inner product on $C^\infty_{\mathbb{C}}(S^2) \supset Poly_{\mathbb{C}}(S^2)_{\leq n}$ (cf. Eq. (4.4)):

$$\langle f_1, f_2 \rangle = \frac{1}{4\pi} \int_{S^2} \overline{f_1} f_2 \, dS = \frac{1}{4\pi} \int_{S^2} \overline{f_1(\mathbf{n})} f_2(\mathbf{n}) d\mathbf{n}, \qquad (7.58)$$

- The induced inner product on $Poly_{\mathbb{C}}(S^2)_{\leq n} \subset C^\infty_{\mathbb{C}}(S^2)$ (cf. Eq. (7.3)):

$$\langle f_1, f_2 \rangle_{\star^n_{\frac{1}{b}}} = \frac{1}{4\pi} \int_{S^2} \overline{f_1} \star^n_{\frac{1}{b}} f_2 \, dS = \frac{1}{4\pi} \int_{S^2} \overline{f_1(\mathbf{n})} \star^n_{\frac{1}{b}} f_2(\mathbf{n}) d\mathbf{n}. \qquad (7.59)$$

Associated with each of these two inner product on $Poly_{\mathbb{C}}(S^2)_{\leq n}$ defined by integration, one can define an integral trikernel $\mathbb{L}$, cf. (7.33). The first one is the Berezin trikernel $\mathbb{L}^j_{\frac{1}{b}}$, also defined implicitly by (cf. Eq. (7.39)):

$$f_1 \star^n_{\frac{1}{b}} f_2 \, (\mathbf{n}) = \iint_{S^2 \times S^2} f_1(\mathbf{n}_1) f_2(\mathbf{n}_2) \mathbb{L}^j_{\frac{1}{b}}(\mathbf{n}_1, \mathbf{n}_2, \mathbf{n}) d\mathbf{n}_1 d\mathbf{n}_2. \qquad (7.60)$$

Definition 7.2.15. *Wildberger's recursive trikernel* is the function

$$\mathbb{T}^j_{\frac{1}{b}} : S^2 \times S^2 \times S^2 \to \mathbb{C}$$

defined implicitly by

$$f_1 \star^n_{\frac{1}{b}} f_2 \, (\mathbf{n}) = \int_{S^2} \left\{ \int_{S^2} \mathbb{T}^j_{\frac{1}{b}}(\mathbf{n}_1, \mathbf{n}_2, \mathbf{n}) \star^n_{\frac{1}{b}} f_1(\mathbf{n}_1) d\mathbf{n}_1 \right\} \star^n_{\frac{1}{b}} f_2(\mathbf{n}_2) d\mathbf{n}_2 \qquad (7.61)$$

where each twisted product $\star^n_{\frac{1}{b}}$ on the r.h.s. of the above equation is taken with respect to the variables being integrated.

One obtains, of course, an explicit expression for the Berezin trikernel $\mathbb{L}^j_{\frac{1}{b}}$ from the Eqs. (7.40)–(7.11). Similarly, Wildberger's trikernel also has such an explicit expression.

Proposition 7.2.16. *Wildberger's recursive trikernel is given by:*

$$\mathbb{T}_{b}^{j}(\mathbf{n}_1, \mathbf{n}_2, \mathbf{n}_3) \tag{7.62}$$

$$= \frac{(-1)^n \sqrt{n+1}}{(4\pi)^2} \sum_{l_i,m_i} \begin{bmatrix} l_1 & l_2 & l_3 \\ m_1 & m_2 & m_3 \end{bmatrix} [j]\, b_{l_1}^n b_{l_2}^n b_{l_3}^n\, \overline{Y_{l_1}^{m_1}}(\mathbf{n}_1) \overline{Y_{l_2}^{m_2}}(\mathbf{n}_2) \overline{Y_{l_3}^{m_3}}(\mathbf{n}_3)$$

where b_l^n are the characteristic numbers of the standard Berezin correspondence, given by Eq. (6.37).

Proof. From Eq. (7.61), we reason analogously to the proof of Corollary 7.2.2, with the difference that, for the induced inner product (7.59), the orthonormal basis vectors are not Y_l^m, but $b_l^n Y_l^m$, with b_l^n given by (6.37).   □

Wildberger's recursive trikernel can be generalized to a recursive trikernel for any symbol correspondence by obvious analogy to the Berezin case:

Corollary 7.2.17. *Let $W_{\hat{c}}^{j}$ be the symbol correspondence given by characteristic numbers $c_l^n \in \mathbb{R}^*$. Its recursive integral trikernel is the polynomial function $\mathbb{T}_{\hat{c}}^{j} \in (Poly_{\mathbb{C}}(S^2)_{\leq n})^3$ given by*

$$\mathbb{T}_{\hat{c}}^{j}(\mathbf{n}_1, \mathbf{n}_2, \mathbf{n}_3) \tag{7.63}$$

$$= \frac{(-1)^n \sqrt{n+1}}{(4\pi)^2} \sum_{l_i,m_i} \begin{bmatrix} l_1 & l_2 & l_3 \\ m_1 & m_2 & m_3 \end{bmatrix} [j]\, c_{l_1}^n c_{l_2}^n c_{l_3}^n\, \overline{Y_{l_1}^{m_1}}(\mathbf{n}_1) \overline{Y_{l_2}^{m_2}}(\mathbf{n}_2) \overline{Y_{l_3}^{m_3}}(\mathbf{n}_3)$$

with the usual (7.37) restrictions on the l_k, m_k summations. It is symmetric in the sense of (7.38) and is implicitly defined by

$$f_1 \star_{\hat{c}}^{n} f_2\, (\mathbf{n}) = \int_{S^2} \left\{ \int_{S^2} \mathbb{T}_{\hat{c}}^{j}(\mathbf{n}_1, \mathbf{n}_2, \mathbf{n}) \star_{\hat{c}}^{n} f_1(\mathbf{n}_1) d\mathbf{n}_1 \right\} \star_{\hat{c}}^{n} f_2(\mathbf{n}_2) d\mathbf{n}_2, \tag{7.64}$$

where each twisted product $\star_{\hat{c}}^{n}$ on the r.h.s. of the above equation is taken with respect to the variables being integrated.

Every recursive trikernel can be written more explicitly, by performing the following substitution into Eq. (7.41),

$$\frac{c_{l_3}^n}{c_{l_1}^n c_{l_2}^n} \rightarrow c_{l_1}^n c_{l_2}^n c_{l_3}^n \,. \tag{7.65}$$

Or it can be rewritten in terms of $SO(3)$-invariant functions, by performing the substitution (7.65) into Eq. (7.51). For Wildberger's recursive trikernel, the above substitution to be performed in Eqs. (7.41) and (7.51) reads:

$$\frac{c_{l_3}^n}{c_{l_1}^n c_{l_2}^n} \rightarrow b_{l_1}^n b_{l_2}^n b_{l_3}^n = (n!\sqrt{n+1})^3 \prod_{k=1}^{3} \frac{1}{\sqrt{(n+l_k+1)!(n-l_k)!}} \,. \tag{7.66}$$

Next, we will express the integrand in (7.64) using only ordinary (commutative) multiplication of functions, that is, no twisted product involved. To this end, we shall invoke the duality $f \longleftrightarrow \tilde{f}$ between symbols in $Poly_{\mathbb{C}}(S^2)_{\leq n}$, with reference to Remark 6.2.25 and Eq. (7.29).

Lemma 7.2.18. *If $f, g \in Poly_{\mathbb{C}}(S^2)_{\leq n}$, then*

$$\int_{S^2} f(\mathbf{n}) \star_{\tilde{c}}^n g(\mathbf{n}) d\mathbf{n} = \int_{S^2} f(\mathbf{n}) \tilde{g}(\mathbf{n}) d\mathbf{n} = \int_{S^2} \tilde{f}(\mathbf{n}) g(\mathbf{n}) d\mathbf{n},$$

where $\tilde{f}$ denotes the contravariant dual of f with respect to $W_{\tilde{c}}^j$, that is,

$$\tilde{f} = U_{\tilde{c},\frac{1}{\tilde{c}}}^j(f), \ \ or \ \ f = U_{\frac{1}{\tilde{c}},\tilde{c}}^j(\tilde{f}) \,, \tag{7.67}$$

cf. Definition 7.1.18.

Proof. Assume $P, Q \in M_{\mathbb{C}}(n+1)$ correspond to f, g via $W_{\tilde{c}}^j$, respectively, and hence by (6.25)

$$\int_{S^2} f(\mathbf{n}) \star_{\tilde{c}}^n g(\mathbf{n}) d\mathbf{n} = \int_{S^2} W_{\tilde{c}}(PQ) dS = \frac{1}{n+1} trace(PQ) = \frac{1}{n+1} trace(QP)$$

$$= \int_{S^2} \widetilde{W_{\tilde{c}}}(Q) W_{\tilde{c}}(P) dS = \int_{S^2} W_{\frac{1}{\tilde{c}}}(Q) W_{\tilde{c}}(P) dS$$

$$= \int_{S^2} \tilde{g}(\mathbf{n}) f(\mathbf{n}) d\mathbf{n} = \int_{S^2} \tilde{f}(\mathbf{n}) g(\mathbf{n}) d\mathbf{n}.$$

$\square$

By applying this lemma twice, the recursive integral equation (7.64) can also be written as

$$f_1 \star_{\tilde{c}}^n f_2 (\mathbf{n}) = \iint_{S^2 \times S^2} \mathbb{T}_{\tilde{c}}^j(\mathbf{n}_1, \mathbf{n}_2, \mathbf{n}) \tilde{f}_1(\mathbf{n}_1) \tilde{f}_2(\mathbf{n}_2) d\mathbf{n}_1 d\mathbf{n}_2 \,. \tag{7.68}$$

Transition Kernels

On the other hand, we shall also make special use of the transition kernel

$$\mathbb{U}_{\tilde{c},\frac{1}{\tilde{c}}}^j : S^2 \times S^2 \rightarrow \mathbb{C}$$

which is associated with the integral form of the transformation $f \to \tilde{f}$ (cf. (7.67)), as follows:

Proposition 7.2.19. *The duality transformations (7.67) can be written in integral form as*

$$\tilde{f}(\mathbf{n}_1) = \int_{S^2} \mathbb{U}^j_{\vec{c},\frac{1}{\vec{c}}}(\mathbf{n}_1, \mathbf{n}_2) f(\mathbf{n}_2) d\mathbf{n}_2 , \tag{7.69}$$

$$f(\mathbf{n}_1) = \int_{S^2} \mathbb{U}^j_{\frac{1}{\vec{c}},\vec{c}}(\mathbf{n}_1, \mathbf{n}_2) \tilde{f}(\mathbf{n}_2) d\mathbf{n}_2 \tag{7.70}$$

with the following transition kernels:

$$\mathbb{U}^j_{\vec{c},\frac{1}{\vec{c}}}(\mathbf{n}_1, \mathbf{n}_2) = \frac{1}{4\pi} \sum_{l=0}^{2j} \frac{2l+1}{(c_l^n)^2} P_l(\mathbf{n}_1 \cdot \mathbf{n}_2) , \tag{7.71}$$

$$\mathbb{U}^j_{\frac{1}{\vec{c}},\vec{c}}(\mathbf{n}_1, \mathbf{n}_2) = \frac{1}{4\pi} \sum_{l=0}^{2j} (c_l^n)^2 (2l+1) P_l(\mathbf{n}_1 \cdot \mathbf{n}_2) . \tag{7.72}$$

Proof. Let $P \in M_{\mathbb{C}}(n+1)$, $P = \sum_{l=0}^n P_l$, where each $P_l \in M_{\mathbb{C}}(\varphi_l)$ decomposes as $P_l = \sum_{m=-l}^l p_{lm} e(l, m)$. Let $f \in Poly_{\mathbb{C}}(S^2)_{\leq n}$ be the symbol of P via the correspondence determined by characteristic numbers $\vec{c}$. Then, $f = \sum_{l=0}^n f_l$, where each $f_l \in Poly(\varphi_l)$ decomposes as $f_l = \sum_{m=-l}^l \frac{1}{\sqrt{n+1}} c_l^n p_{lm} Y_{lm}$.

It now follows straightforwardly from (7.44)–(7.45), (7.69), and (7.71) that $\tilde{f} = \sum_{l=0}^n \tilde{f}_l$, where $\tilde{f}_l = \sum_{m=-l}^l \frac{1}{\sqrt{n+1}} \frac{p_{lm}}{c_l^n} Y_{lm}$. Therefore, $\tilde{f}$ is the contravariant dual of f. Similarly for (7.70) and (7.72). $\qquad\square$

Note that in the above proof, we may as well replace $\frac{1}{\vec{c}}$ by any other string $\vec{c}'$ of characteristic numbers. In other words, by similar reasoning we generalize Eqs. (7.69) and (7.70), as follows:

Proposition 7.2.20. *The transition operator* $U^j_{\vec{c},\vec{c}'} = W^j_{\vec{c}'} \circ (W^j_{\vec{c}})^{-1}$, *cf. Eq. (7.29), which takes the symbol* $f = W^j_{\vec{c}}[P]$ *of P in correspondence defined by $\vec{c}$ to the symbol* $f' = W^j_{\vec{c}'}[P]$ *in correspondence defined by $\vec{c}'$, can be written in integral form as*

$$f'(\mathbf{n}_1) = \int_{S^2} \mathbb{U}^j_{\vec{c},\vec{c}'}(\mathbf{n}_1, \mathbf{n}_2) f(\mathbf{n}_2) d\mathbf{n}_2 , \tag{7.73}$$

where

$$\mathbb{U}^j_{\vec{c},\vec{c}'}(\mathbf{n}_1, \mathbf{n}_2) = \frac{1}{4\pi} \sum_{l=0}^{2j} \frac{(c_l^n)'}{c_l^n} (2l+1) P_l(\mathbf{n}_1 \cdot \mathbf{n}_2) . \tag{7.74}$$

Moreover, for any pair $(\vec{c}, \vec{c}')$ *the* transition kernels $\mathbb{U}^j_{\vec{c},\vec{c}'}$ *and* $\mathbb{U}^j_{\vec{c}',\vec{c}}$ *yield the reproducing kernel by the formula*

$$R^j(\mathbf{n}_1, \mathbf{n}_2) = \int_{S^2} \mathbb{U}^j_{\vec{c},\vec{c}'}(\mathbf{n}_1, \mathbf{n}) \mathbb{U}^j_{\vec{c}',\vec{c}}(\mathbf{n}, \mathbf{n}_2) d\mathbf{n}, \text{ cf. (7.44)–(7.45).} \tag{7.75}$$

Furthermore, the composition of transition operators $\mathbb{U}^j_{\vec{c},\vec{c}'}$ *yields the following "composition" rule for transition kernels*

$$\int_{S^2} \mathbb{U}^j_{\vec{c},\vec{c}'}(\mathbf{n}_1, \mathbf{n}_2) \mathbb{U}^j_{\vec{c}',\vec{c}''}(\mathbf{n}_2, \mathbf{n}_3) d\mathbf{n}_2 = \mathbb{U}^j_{\vec{c},\vec{c}''}(\mathbf{n}_1, \mathbf{n}_3). \tag{7.76}$$

Remark 7.2.21. Note that the reproducing kernel $R^j(\mathbf{n}_1, \mathbf{n}_2)$ is an $SO(3)$-invariant function on $S^2 \times S^2$ which decomposes as

$$R^j(\mathbf{n}_1, \mathbf{n}_2) = \sum_{l=0}^{n} R^j_l(\mathbf{n}_1, \mathbf{n}_2), \quad R^j_l(\mathbf{n}_1, \mathbf{n}_2) = \frac{2l+1}{4\pi} P_l(\mathbf{n}_1 \cdot \mathbf{n}_2), \tag{7.77}$$

so that for every symbol $f \in Poly_{\mathbb{C}}(S^2)_{\leq n}$ which is decomposed into the l-invariant subspaces as $f = \sum_{l=0}^{n} f_l$, we have

$$\int_{S^2} f(\mathbf{n}_1) R^j_l(\mathbf{n}_1, \mathbf{n}_2) d\mathbf{n}_1 = f_l(\mathbf{n}_2). \tag{7.78}$$

Combining Eqs. (7.68)–(7.70), we obtain

Proposition 7.2.22. *The bona-fide and the recursive integral trikernels for a twisted product* $\star^n_{\vec{c}}$ *are related by the integral equations*

$$\mathbb{L}^j_{\vec{c}}(\mathbf{n}_1, \mathbf{n}_2, \mathbf{n}) = \int_{S^2 \times S^2} \mathbb{U}^j_{\vec{c},\frac{1}{\vec{c}}}(\mathbf{n}_1, \mathbf{n}'_1) \mathbb{U}^j_{\vec{c},\frac{1}{\vec{c}}}(\mathbf{n}_2, \mathbf{n}'_2) \mathbb{T}^j_{\vec{c}}(\mathbf{n}'_1, \mathbf{n}'_2, \mathbf{n}) d\mathbf{n}'_1 d\mathbf{n}'_2, \tag{7.79}$$

$$\mathbb{T}^j_{\vec{c}}(\mathbf{n}_1, \mathbf{n}_2, \mathbf{n}) = \int_{S^2 \times S^2} \mathbb{U}^j_{\frac{1}{\vec{c}},\vec{c}}(\mathbf{n}_1, \mathbf{n}'_1) \mathbb{U}^j_{\frac{1}{\vec{c}},\vec{c}}(\mathbf{n}_2, \mathbf{n}'_2) \mathbb{L}^j_{\vec{c}}(\mathbf{n}'_1, \mathbf{n}'_2, \mathbf{n}) d\mathbf{n}'_1 d\mathbf{n}'_2. \tag{7.80}$$

The following proposition is obtained straightforwardly from the above Eq. (7.74) for the transition kernels $\mathbb{U}^j_{\vec{c},\vec{c}'}$, see also Eq. (7.46), and the general formula (7.40) for the twisted product integral trikernels $\mathbb{L}^j_{\vec{c}}$.

Proposition 7.2.23. *The bona-fide integral trikernels of two distinct twisted products* $\star^n_{\vec{c}}$ *and* $\star^n_{\vec{c}'}$ *are related by*

$$\mathbb{L}^j_{\vec{c}}(\mathbf{n}_1, \mathbf{n}_2, \mathbf{n}_3) \tag{7.81}$$

$$= \int_{S^2 \times S^2 \times S^2} \mathbb{U}^j_{\vec{c},\vec{c}'}(\mathbf{n}_1, \mathbf{n}'_1) \mathbb{U}^j_{\vec{c},\vec{c}'}(\mathbf{n}_2, \mathbf{n}'_2) \mathbb{U}^j_{\vec{c}',\vec{c}}(\mathbf{n}_3, \mathbf{n}'_3) \mathbb{L}^j_{\vec{c}'}(\mathbf{n}'_1, \mathbf{n}'_2, \mathbf{n}'_3) d\mathbf{n}'_1 d\mathbf{n}'_2 d\mathbf{n}'_3,$$

Similarly for the relation between recursive trikernels of different symbol correspondences and also for the relation between the bona-fide trikernel of one correspondence and the recursive trikernel of another correspondence, by appropriate use of transition kernels (7.74) and (7.71)–(7.72).

Duality of Twisted Products

In this way it is easy to see that, for any non-self-dual symbol correspondence, with its twisted product $\star^n_{\overset{1}{c}}$, the covariant-contravariant duality is not a twisted product homomorphism, but instead, we have the following

Corollary 7.2.24. *The covariant-contravariant duality satisfies*

$$\widetilde{f \star^n_{\overset{}{c}} g} = \tilde{f} \star^n_{\overset{1}{\overset{}{c}}} \tilde{g} . \tag{7.82}$$

Proof. This follows straightforwardly from Eqs. (7.68)–(7.81). □

Definition 7.2.25. In view of Eq. (7.82) above, the twisted product $\star^n_{\overset{1}{\overset{}{c}}}$ shall be called the *dual twisted product* of $\star^n_{\overset{}{c}}$, also denoted $\tilde{\star}^n_{\overset{}{c}}$. The dual of the standard Berezin twisted product shall be called the *standard Toeplitz twisted product*, denoted by $\star^n_{\overset{1}{\overset{}{b}}}$, and similarly for the dual of the alternate Berezin twisted product.

Remark 7.2.26. For a given symbol correspondence with characteristic numbers $\vec{c}$, the twisted product $\star^n_{\overset{}{c}}$ is the natural product of covariant symbols, while the dual twisted product $\star^n_{\overset{1}{\overset{}{c}}}$ is the natural product of contravariant symbols, which are induced from the operator product. In the literature, the association of an operator to a function via the contravariant Berezin symbol correspondence (as given by Eq. (6.24), but see also Theorem 6.2.21 and Remark 6.2.25) is known as Toeplitz quantization (see, e.g. [18, 21] and references therein).

Definition 7.2.27. According to Definition 7.2.25 and Remark 7.2.26 above, the symbol correspondence with characteristic numbers $\frac{1}{b}$ shall be called the *standard Toeplitz symbol correspondence* and the one with characteristic numbers $\frac{1}{b-}$ shall be called the *alternate Toeplitz symbol correspondence*.

Explicit expressions for the standard Toeplitz twisted product of spherical harmonics and the standard Toeplitz integral trikernel are obtained from Eqs. (7.9), (7.39), (7.41) and (7.50)–(7.51), via the following substitution:

$$\frac{c^n_{l_3}}{c^n_{l_1} c^n_{l_2}} \rightarrow \frac{b^n_{l_1} b^n_{l_2}}{b^n_{l_3}} = \sqrt{\frac{\binom{n}{l_1}\binom{n}{l_2}\binom{n+l_3+1}{l_3}}{\binom{n+l_1+1}{l_1}\binom{n+l_2+1}{l_2}\binom{n}{l_3}}} \tag{7.83}$$

$$= \sqrt{\frac{(n+l_3+1)!(n-l_3)!(n+1)!n!}{(n+l_1+1)!(n+l_2+1)!(n-l_1)!(n-l_2)!}} .$$

7.2.3 Other Formulae Related to Integral Trikernels

Integral Formulae for Trikernels

Returning to Wildberger's recursive trikernel (i.e., for the standard Berezin correspondence), its nicest feature is its simple closed formula, presented in the following proposition, whose proof is delegated to Appendix 10.8.

Proposition 7.2.28 ([92]). *Recalling the notation from Lemma 7.2.9, Wildberger's recursive trikernel is the function* $\mathbb{T}_{\overset{j}{b}} : S^2 \times S^2 \times S^2 \to \mathbb{C}$ *given by*

$$\mathbb{T}_{\overset{j}{b}}(\mathbf{n}_1, \mathbf{n}_2, \mathbf{n}_3) = \left(\frac{n+1}{2^n 4\pi}\right)^2 \left(1 + X(\mathbf{n}_1, \mathbf{n}_2, \mathbf{n}_3) + i\,[\mathbf{n}_1, \mathbf{n}_2, \mathbf{n}_3]\right)^n \qquad (7.84)$$

where $i = \sqrt{-1}$ *and* $X(\mathbf{n}_1, \mathbf{n}_2, \mathbf{n}_3)$ *is defined by (7.53) and* $[\mathbf{n}_1, \mathbf{n}_2, \mathbf{n}_3]$ *is the* 3×3 *determinant. One should compare the simple closed formula (7.84) with the formulae that are obtained from (7.41) and (7.51) via substitution (7.66).*

Then, an integral equation for general bona-fide trikernels is obtained from Wildberger's recursive trikernel (7.84) and Eqs. (6.37) and (7.69)–(7.81).

Theorem 7.2.29. *The* $\vec{c}$-*correspondence trikernel* $\mathbb{L}_{\overset{j}{c}}$ *is itself also expressed by the following sum of integrals:*

$$\mathbb{L}_{\overset{j}{c}}(\mathbf{n}_1, \mathbf{n}_2, \mathbf{n}_3) = \left(\frac{n+1}{2^n 4\pi}\right)^2 \sum_{l_1,l_2,l_3=0}^{n} \frac{c_{l_3}^n}{c_{l_1}^n c_{l_2}^n} \frac{1}{b_{l_1}^n b_{l_2}^n b_{l_3}^n} \mathcal{I}_{l_1,l_2,l_3}^n(\mathbf{n}_1, \mathbf{n}_2, \mathbf{n}_3)$$

$$= \frac{\sqrt{n+1}}{(n!)^3 4^n (4\pi)^2} \sum_{l_1,l_2,l_3=0}^{n} \frac{c_{l_3}^n}{c_{l_1}^n c_{l_2}^n} \prod_{k=1}^{3} \sqrt{(n+l_k+1)!(n-l_k)!} \; \mathcal{I}_{l_1,l_2,l_3}^n(\mathbf{n}_1, \mathbf{n}_2, \mathbf{n}_3)$$

$$(7.85)$$

where

$$\mathcal{I}_{l_1,l_2,l_3}^n(\mathbf{n}_1, \mathbf{n}_2, \mathbf{n}_3) = \frac{1}{(4\pi)^3} \int_{S^2 \times S^2 \times S^2} \prod_{k=1}^{3} (2l_k + 1) P_{l_k}(\mathbf{n}_k \cdot \mathbf{n}_k') \qquad (7.86)$$

$$\cdot \left(1 + X(\mathbf{n}_1', \mathbf{n}_2', \mathbf{n}_3') + i\,[\mathbf{n}_1', \mathbf{n}_2', \mathbf{n}_3']\right)^n d\mathbf{n}_1' d\mathbf{n}_2' d\mathbf{n}_3'.$$

The standard Stratonovich trikernel is obtained from the above formula by setting all $c_l^n = 1$, whereas for the standard Berezin case, $c_l^n = b_l^n$ is given by (6.37), so that

$$\frac{c_{l_3}^n}{c_{l_1}^n c_{l_2}^n} \frac{1}{b_{l_1}^n b_{l_2}^n b_{l_3}^n} = \frac{1}{(b_{l_1}^n)^2 (b_{l_2}^n)^2} = \frac{(n+l_1+1)!(n-l_1)!(n+l_2+1)!(n-l_2)!}{(n+1)^2 (n!)^4}$$

and the summation in l_3 yields the reproducing kernel (cf. Eq. (7.45)), which eliminates one integration on S^2. Thus, the standard Berezin trikernel is given by

$$\mathbb{L}_b^j(\mathbf{n}_1, \mathbf{n}_2, \mathbf{n}_3) \tag{7.87}$$

$$= \frac{1}{(n!)^4 4^n (4\pi)^4} \sum_{l_1,l_2=0}^{n} (n + l_1 + 1)!(n-l_1)!(2l_1 + 1)(n + l_2 + 1)!(n - l_2)!(2l_2 + 1)$$

$$\cdot \int_{S^2 \times S^2} P_{l_1}(\mathbf{n}_1 \cdot \mathbf{n}_1') P_{l_2}(\mathbf{n}_2 \cdot \mathbf{n}_2') \Big(1 + X(\mathbf{n}_1', \mathbf{n}_2', \mathbf{n}_3) + i [\mathbf{n}_1', \mathbf{n}_2', \mathbf{n}_3] \Big)^n d\mathbf{n}_1' d\mathbf{n}_2' .$$

On the other hand, the integral trikernel for the standard Toeplitz twisted product (cf. Definition 7.2.25) is obtained from (7.85) by setting $c_l^n = \frac{1}{b_l^n}$, so that

$$\frac{c_{l_3}^n}{c_{l_1}^n c_{l_2}^n} \frac{1}{b_{l_1}^n b_{l_2}^n b_{l_3}^n} = \frac{1}{(b_{l_3}^n)^2} = \frac{(n + l_3 + 1)!(n - l_3)!}{(n + 1)(n!)^2}$$

and now both summations in l_1 and l_2 yield reproducing kernels that eliminate two integrations on S^2. Therefore, the standard Toeplitz trikernel is given by

$$\mathbb{L}_b^j(\mathbf{n}_1, \mathbf{n}_2, \mathbf{n}_3) = \frac{n + 1}{(n!)^2 4^n (4\pi)^3} \sum_{l_3=0}^{n} (n + l_3 + 1)!(n - l_3)!(2l_3 + 1)$$

$$\cdot \int_{S^2} P_{l_3}(\mathbf{n}_3 \cdot \mathbf{n}_3') \Big(1 + X(\mathbf{n}_1, \mathbf{n}_2, \mathbf{n}_3') + i [\mathbf{n}_1, \mathbf{n}_2, \mathbf{n}_3'] \Big)^n d\mathbf{n}_3' . \tag{7.88}$$

Special Functional Transforms

Note that formulae (7.87)–(7.88) could have been obtained directly from Eqs. (7.69)–(7.79). Note also that, in these two formulae above, we have used the covariant-to-contravariant transition kernel for the standard Berezin symbol correspondence, namely

$$\mathbb{U}_{b, \frac{1}{b}}^j(\mathbf{n}, \mathbf{n}') = \frac{1}{4\pi} \sum_{l=0}^{n} \frac{\binom{n+l+1}{l}}{\binom{n}{l}} (2l + 1) P_l(\mathbf{n} \cdot \mathbf{n}'). \tag{7.89}$$

Its inverse transition kernel has a simple closed formula, as follows:

Proposition 7.2.30. *The contravariant-to-covariant transition kernel for the standard Berezin symbol correspondence is given by*

$$U^j_{\frac{1}{b},\vec{b}}(\mathbf{n},\mathbf{n}') = \frac{1}{4\pi}\sum_{l=0}^{n}\frac{\binom{n}{l}}{\binom{n+l+1}{l}}(2l+1)P_l(\mathbf{n}\cdot\mathbf{n}') = \frac{n+1}{4\pi}\left(\frac{1+\mathbf{n}\cdot\mathbf{n}'}{2}\right)^n.$$

(7.90)

Remark 7.2.31.   (i) Formula (7.90) can be readily obtained from [11, 12], but in Appendix 10.9 we present a proof of this formula following more closely to [92].

(ii) Note that Eq. (7.90) is the "inverse" of Eq. (4.36), that is,

$$\left(\frac{1+z}{2}\right)^n = \sum_{l=0}^{n}\frac{\binom{n}{l}}{\binom{n+l+1}{l}}\frac{2l+1}{n+1}P_l(z) = (n!)^2\sum_{l=0}^{n}\frac{(2l+1)P_l(z)}{(n+l+1)!(n-l)!}.$$

(7.91)

(iii) However, a simple closed formula for $U^j_{\vec{b},\frac{1}{b}}(\mathbf{n},\mathbf{n}')$ remains unknown to us.

Similarly, a simple closed formula like (7.84) for general values of j is not yet known for the Berezin trikernel or even for the Toeplitz trikernel. The situation is the same for the Stratonovich trikernel.

Definition 7.2.32. Due to the simple closed form for $U^j_{\frac{1}{b},\vec{b}}$, the integral equation

$$f(\mathbf{n}) = \frac{n+1}{4\pi}\int_{S^2}\left(\frac{1+\mathbf{n}\cdot\mathbf{n}'}{2}\right)^n\tilde{f}(\mathbf{n}')d\mathbf{n}' =: \mathcal{B}[\tilde{f}](\mathbf{n})\qquad(7.92)$$

is known as the *Berezin transform* of a function $\tilde{f}$ on S^2.

In view of (iii) in the above remark, no such simple expression as (7.92) is known to us for the inverse transform $f \to \tilde{f}$. Nonetheless, it follows from (7.89) that the *inverse Berezin transform* of a function f on S^2 is given by

$$\mathcal{B}^{-1}[f](\mathbf{n}) = \tilde{f}(\mathbf{n}) = \frac{1}{(n!)^2(n+1)}\sum_{l=0}^{n}(n+l+1)!(n-l)!\frac{(2l+1)}{4\pi}\int_{S^2}P_l(\mathbf{n}\cdot\mathbf{n}')f(\mathbf{n}')d\mathbf{n}'$$

(7.93)

Remark 7.2.33.   (i) The reader should be aware that the term "Berezin transform" of an operator is also sometimes used in the literature to indicate the standard Berezin (covariant) symbol of an operator. Also, the Berezin transform of a function on S^2 is more commonly presented using holomorphic coordinates on $\mathbb{C}$, which is identified with S^2 via the stereographic projection (2.35), as was originally done by Berezin [11, 12].

(ii) Clearly, both the Berezin transform of a function on S^2, $\mathcal{B}$ defined by (7.92), and its inverse, $\mathcal{B}^{-1}$ defined by (7.93), depend on n. Also, they both vanish on the complement of $Poly_{\mathbb{C}}(S^2)_{\leq n}$, so that composing the Berezin transform with its inverse (or vice-versa) amounts to orthogonally projecting any function on

S^2 to the subspace $Poly_\mathbb{C}(S^2)_{\leq n}$. In fact, this follows as a particular case of Eq. (7.75), for $\vec{c} = \vec{b}$, $\vec{c}' = \frac{1}{\vec{b}}$.

In the category of characteristic-positive symbol correspondences, the standard Stratonovich symbol correspondence is naturally singled out, standing in a prominent position together with its twisted product. For this reason, the standard Stratonovich trikernel is also singled out. We also observe from Theorem 7.2.29 that the equation which yields the standard Stratonovich trikernel from Wildberger's recursive trikernel can be written more simply as follows:

$$\mathbb{L}_1^j(\mathbf{n}_1, \mathbf{n}_2, \mathbf{n}_3) = \tag{7.94}$$

$$\int_{S^2 \times S^2 \times S^2} \mathbb{U}_{\vec{b},1}^j(\mathbf{n}_1, \mathbf{n}_1') \mathbb{U}_{\vec{b},1}^j(\mathbf{n}_2, \mathbf{n}_2') \mathbb{U}_{\vec{b},1}^j(\mathbf{n}_3, \mathbf{n}_3') \mathbb{T}_{\vec{b}}^j(\mathbf{n}_1', \mathbf{n}_2', \mathbf{n}_3') d\mathbf{n}_1' d\mathbf{n}_2' d\mathbf{n}_3'$$

and this, in turn, singles out the importance of the transition kernel $\mathbb{U}_{\vec{b},1}^j(\mathbf{n}, \mathbf{n}')$.

Definition 7.2.34. The *Berezin-Stratonovich transition kernel*

$$\mathbb{U}_{\vec{b},1}^j(\mathbf{n}, \mathbf{n}') = \frac{1}{4\pi} \sum_{l=0}^n \frac{2l+1}{b_l^n} P_l(\mathbf{n} \cdot \mathbf{n}')$$

defines the *Berezin-Stratonovich transform*

$$\mathcal{BS} : Poly_\mathbb{C}(S^2)_{\leq n} \to Poly_\mathbb{C}(S^2)_{\leq n}$$

via the integral equation

$$\mathcal{BS}[f](\mathbf{n}) = \frac{1}{n!\sqrt{n+1}} \sum_{l=0}^n \sqrt{(n+l+1)!(n-l)!} \frac{(2l+1)}{4\pi} \int_{S^2} P_l(\mathbf{n}\cdot\mathbf{n}') f(\mathbf{n}')d\mathbf{n}'. \tag{7.95}$$

Its inverse, the *Stratonovich-Berezin transform* $\mathcal{SB} = (\mathcal{BS})^{-1}$ is given by

$$\mathcal{SB}[f](\mathbf{n}) = n!\sqrt{n+1} \sum_{l=0}^n \frac{1}{\sqrt{(n+l+1)!(n-l)!}} \frac{(2l+1)}{4\pi} \int_{S^2} P_l(\mathbf{n}\cdot\mathbf{n}') f(\mathbf{n}')d\mathbf{n}'. \tag{7.96}$$

The following lemma, whose straightforward proof is analogous to the proof of the equations in Proposition 7.2.20, gives the relation between the two special transforms on $Poly_\mathbb{C}(S^2)_{\leq n}$ defined above.

Lemma 7.2.35. *The Berezin transform is the square of the Stratonovich-Berezin transform,*

$$\mathcal{B} = (\mathcal{SB})^2 = \mathcal{SB} \circ \mathcal{SB}. \tag{7.97}$$

Remark 7.2.36. One should note, however, that although $\mathcal{SB}$ is the unique *positive* square root of $\mathcal{B}$, there are $(\mathbb{Z}_2)^n$ distinct square roots of $\mathcal{B}$, one for each choice of Stratonovich correspondence with characteristic numbers $\vec{\varepsilon} = (\varepsilon_l^n)$, $\varepsilon_l^n = \pm 1$, defining transition kernels $\mathbb{U}_{\vec{b},\vec{\varepsilon}}^j$ and $\mathbb{U}_{\vec{\varepsilon},\vec{b}}^j$.

Again, just as for $\mathcal{B}^{-1}$, simple closed formulae like (7.92) for either $\mathcal{SB}$ or $\mathcal{BS}$ are yet unknown to us.

Integral Formulae for Twisted Products of Spherical Harmonics

Now, for symbols $f, g \in Poly_{\mathbb{C}}(S^2)_{\leq n}$ which are decomposed into the l-invariant subspaces as $f = \sum_{l=0}^n f_l$, $g = \sum_{l=0}^n g_l$, we have from Remark 7.2.21 (cf. Eqs. (7.77)–(7.78)) and Eqs. (7.39) and (7.85)–(7.86), that

$$f \star_{\vec{c}}^n g(\mathbf{n}_3) = \left(\frac{n+1}{2^n 4\pi}\right)^2 \sum_{l_1,l_2,l_3=0}^n \frac{c_{l_3}^n}{c_{l_1}^n c_{l_2}^n} \frac{1}{b_{l_1}^n b_{l_2}^n b_{l_3}^n} \frac{2l_3+1}{4\pi} \tag{7.98}$$

$$\cdot \int_{S^2 \times S^2 \times S^2} f_{l_1}(\mathbf{n}_1') g_{l_2}(\mathbf{n}_2') P_{l_3}(\mathbf{n}_3 \cdot \mathbf{n}_3') \left(1 + X(\mathbf{n}_1', \mathbf{n}_2', \mathbf{n}_3') + i[\mathbf{n}_1', \mathbf{n}_2', \mathbf{n}_3']\right)^n d\mathbf{n}_1' d\mathbf{n}_2' d\mathbf{n}_3'$$

which for the standard Berezin case reduces to

$$f \star_{\vec{b}}^n g(\mathbf{n}_3) = \left(\frac{n+1}{2^n 4\pi}\right)^2 \sum_{l_1,l_2=0}^n \frac{\binom{n+l_1+1}{l_1}\binom{n+l_2+1}{l_2}}{\binom{n}{l_1}\binom{n}{l_2}} \tag{7.99}$$

$$\cdot \int_{S^2 \times S^2} f_{l_1}(\mathbf{n}_1') g_{l_2}(\mathbf{n}_2') \left(1 + X(\mathbf{n}_1', \mathbf{n}_2', \mathbf{n}_3) + i[\mathbf{n}_1', \mathbf{n}_2', \mathbf{n}_3]\right)^n d\mathbf{n}_1' d\mathbf{n}_2'.$$

In particular, Eqs. (7.98) and (7.99) above yield alternative expressions for the twisted products of spherical harmonics.

Corollary 7.2.37. *The standard Berezin twisted product of spherical harmonics can also be written as*

$$Y_{l_1}^{m_1} \star_{\vec{b}}^n Y_{l_2}^{m_2}(\mathbf{n}_3) = \frac{(n+l_1+1)!(n-l_1)!(n+l_2+1)!(n-l_2)!}{(n!)^4 4^n (4\pi)^2} \tag{7.100}$$

$$\cdot \int_{S^2 \times S^2} Y_{l_1}^{m_1}(\mathbf{n}_1') Y_{l_2}^{m_2}(\mathbf{n}_2') \left(1 + X(\mathbf{n}_1', \mathbf{n}_2', \mathbf{n}_3) + i[\mathbf{n}_1', \mathbf{n}_2', \mathbf{n}_3]\right)^n d\mathbf{n}_1' d\mathbf{n}_2'$$

while a general twisted product of spherical harmonics is given by

$$Y_{l_1}^{m_1} \star_{\tilde{c}}^n Y_{l_2}^{m_2}(\mathbf{n}_3) = \left(\frac{n+1}{2^n 4\pi}\right)^2 \frac{1}{c_{l_1}^n c_{l_2}^n} \frac{1}{b_{l_1}^n b_{l_2}^n} \sum_{l_3=0}^{n} \frac{c_{l_3}^n}{b_{l_3}^n} \frac{2l_3+1}{4\pi} \tag{7.101}$$

$$\cdot \int_{S^2 \times S^2 \times S^2} Y_{l_1}^{m_1}(\mathbf{n}_1') Y_{l_2}^{m_2}(\mathbf{n}_2') P_{l_3}(\mathbf{n}_3 \cdot \mathbf{n}_3') \Big(1 + X(\mathbf{n}_1', \mathbf{n}_2', \mathbf{n}_3') + i[\mathbf{n}_1', \mathbf{n}_2', \mathbf{n}_3']\Big)^n d\mathbf{n}_1' d\mathbf{n}_2' d\mathbf{n}_3'$$

from which the standard Stratonovich product is obtained by setting all $c_l^n = 1$, as

$$Y_{l_1}^{m_1} \star_1^n Y_{l_2}^{m_2}(\mathbf{n}_3) \tag{7.102}$$

$$= \frac{\sqrt{n+1}}{(n!)^3 4^n (4\pi)^3}^2 \prod_{k=1}^{2} \sqrt{(n+l_k+1)!(n-l_k)!} \sum_{l_3=0}^{n} \sqrt{(n+l_3+1)!(n-l_3)!}(2l_3+1)$$

$$\cdot \int_{S^2 \times S^2 \times S^2} Y_{l_1}^{m_1}(\mathbf{n}_1') Y_{l_2}^{m_2}(\mathbf{n}_2') P_{l_3}(\mathbf{n}_3 \cdot \mathbf{n}_3') \Big(1 + X(\mathbf{n}_1', \mathbf{n}_2', \mathbf{n}_3') + i[\mathbf{n}_1', \mathbf{n}_2', \mathbf{n}_3']\Big)^n d\mathbf{n}_1' d\mathbf{n}_2' d\mathbf{n}_3'.$$

Relationship with Spherical Geometry

Finally, we note that the above formulae (7.84)–(7.88) and (7.98)–(7.102) can also be rewritten in polar form, in terms of the geometry of the vertex spherical triangle spanned by the triple $(\mathbf{n}_1, \mathbf{n}_2, \mathbf{n}_3)$ of unit vectors. To this end, let us write

$$\cos \alpha_i = \mathbf{n}_j \cdot \mathbf{n}_k, \alpha_i \in [0, \pi], \text{ for } \{i, j, k\} = \{1, 2, 3\},$$

$$\beta_i = \alpha_i / 2 \in [0, \pi/2]$$

and consider the geodesic triangle with vertices $\mathbf{n}_i$ and opposite edges of length α_i, positive orientation given by $\mathbf{n}_i \to \mathbf{n}_{i+1}(i \mod 3)$, and oriented (symplectic) area denoted by Θ :

$$-2\pi \leq \Theta(\mathbf{n}_1, \mathbf{n}_2, \mathbf{n}_3) \leq 2\pi$$

Lemma 7.2.38. *Let $\mathbf{n}_k \in S^2(1), k \in \{1, 2, 3\}$, be points on the standard unit sphere of total (symplectic) oriented area 4π, and consider the geodesic triangle with vertices $\mathbf{n}_k$ and arcs of length $2\beta_k$ and signed area Θ, as explained above. Then*

$$\cos \beta_1 \cos \beta_2 \cos \beta_3 = \frac{1}{4} |1 + X(\mathbf{n}_1, \mathbf{n}_2, \mathbf{n}_3) + i[\mathbf{n}_1, \mathbf{n}_2, \mathbf{n}_3]| \tag{7.103}$$

$$\Theta(\mathbf{n}_1, \mathbf{n}_2, \mathbf{n}_3) = 2 \arg(1 + X(\mathbf{n}_1, \mathbf{n}_2, \mathbf{n}_3) + i[\mathbf{n}_1, \mathbf{n}_2, \mathbf{n}_3]) \tag{7.104}$$

where $X(\mathbf{n}_1, \mathbf{n}_2, \mathbf{n}_3)$ *is given by (7.53) and* $[\mathbf{n}_1, \mathbf{n}_2, \mathbf{n}_3]$ *is the* 3×3 *determinant (cf. (7.49)).*

Proof. The Eq. (7.104) for the area $\Theta(\mathbf{n}_1, \mathbf{n}_2, \mathbf{n}_3)$ of the vertex geodesic triangle on S^2 has long been known (at least since Euler, probably much earlier, see also [62]). To prove Eq. (7.103), we use another known identity (see [61]),

$$[\mathbf{n}_1, \mathbf{n}_2, \mathbf{n}_3]^2 = 1 - \cos^2 \alpha_1 - \cos^2 \alpha_2 - \cos^2 \alpha_3 + 2 \cos \alpha_1 \cos \alpha_2 \cos \alpha_3 , \qquad (7.105)$$

which is readily seen to imply (7.103).   □

Corollary 7.2.39 ([92]). *For every* $n = 2j \in \mathbb{N}$, *Wildberger's recursive trikernel* $\mathbb{T}_{\frac{j}{b}} : S^2 \times S^2 \times S^2 \to \mathbb{C}$, *cf. (7.84), can be written in polar form as*

$$\mathbb{T}_{\frac{j}{b}}(\mathbf{n}_1, \mathbf{n}_2, \mathbf{n}_3) = \left(\frac{n+1}{4\pi}\right)^2 A_W^n(\mathbf{n}_1, \mathbf{n}_2, \mathbf{n}_3) \exp\{i\, \Phi_W^n(\mathbf{n}_1, \mathbf{n}_2, \mathbf{n}_3)\} \qquad (7.106)$$

where the amplitude and phase functions are given respectively by

$$A_W^n(\mathbf{n}_1, \mathbf{n}_2, \mathbf{n}_3) = \cos^n \beta_1 \cos^n \beta_2 \cos^n \beta_3 \qquad (7.107)$$

$$= \frac{1}{2^{3n/2}}[(1 + \mathbf{n}_1 \cdot \mathbf{n}_2)(1 + \mathbf{n}_2 \cdot \mathbf{n}_3)(1 + \mathbf{n}_3 \cdot \mathbf{n}_1)]^{n/2}$$

$$\Phi_W^n(\mathbf{n}_1, \mathbf{n}_2, \mathbf{n}_3) = \frac{n}{2}\Theta(\mathbf{n}_1, \mathbf{n}_2, \mathbf{n}_3). \qquad (7.108)$$

In view of the fact that the standard Stratonovich-Weyl twisted product can be seen as the spin version of the ordinary Moyal-Weyl product, whose integral form obtained by Groenewold and von Neumann [35, 77] can be written in terms of the geometry of triangles described by their midpoints (see e.g. [60, 62]), one could hope that a result similar to Corollary 7.2.39 above would hold for the Stratonovich trikernel, with vertex spherical triangles replaced by midpoint spherical triangles (i.e. spherical triangles described by their midpoints, with its area function). This possibility was first set forth by Weinstein [83] and has been partially investigated by Tuynman in collaboration with one of the authors [62].

However, by comparing with the geometrical formulae for midpoint triangles (area function, etc.) presented in [61, 62], it is clear from Eqs. (7.54)–(7.55) in Example 7.2.14, as well the more general Eqs. (7.50)–(7.51), (7.85)–(7.86), that such a very simple closed "midpoint formula" in the style of Corollary 7.2.39 only has a chance of holding for the Stratonovich trikernel $\mathbb{L}_1^j$ asymptotically, as $n \to \infty$, as in a WKB-style approximation.

But such an asymptotic study of the Stratonovich trikernel lies outside the scope of this monograph. The more basic asymptotic study to be presented in the next chapter does not touch on this matter.

Chapter 8
Beginning Asymptotic Analysis of Twisted Products

As mentioned in Remarks 3.3.13 and 4.1.10, while the quantum dynamics of an operator $F \in \mathcal{B}(\mathcal{H}_j)$ is governed by Heisenberg's equation (3.87), the classical dynamics of a function $f \in C_{\mathbb{C}}^{\infty}(S^2)$ is governed by Hamilton's equation (4.16).

Now, via a symbol correspondence $W_{\tilde{c}}^{j}$, Heisenberg's equation can be transformed into a dynamical equation for the symbol of F, $f \in Poly_{\mathbb{C}}(S^2)_{\leq n} \subset C_{\mathbb{C}}^{\infty}(S^2)$, by substituting the commutator of operators $[H, F]$ by the twisted commutator of symbols, $[h, f]_{\star_{\tilde{c}}^{n}} = h \star_{\tilde{c}}^{n} f - f \star_{\tilde{c}}^{n} h$, where $h \in Poly_{\mathbb{C}}(S^2)_{\leq n}$ is the $W_{\tilde{c}}^{j}$-symbol of the preferred Hamiltonian operator H.

Therefore, a natural question is whether Hamilton's equation, which can be defined via the Poisson bracket, can be obtained from Heisenberg's equation for symbols, which is defined via the $\star_{\tilde{c}}^{n}$ commutator, in a suitable limit.[1] In other words, whether Poisson dynamics emerge from quantum dynamics in a suitable asymptotic limit, the so-called (semi)classical limit.

Historically, this question was first addressed in the context of operators on infinite dimensional Hilbert space $L^2(\mathbb{R}^k)$ and functions on affine symplectic space $\mathbb{R}^{2k}$. In that context of affine mechanical systems, the (semi)classical limit, the limit of very large quantum numbers, can be formally treated as the limit $\hbar \to 0$.

However, in the context of spin systems, the (semi)classical limit, the limit of very large quantum numbers, is the limit $2j = n \to \infty$ and, in this context, $\hbar$ is best treated as a constant which can be omitted by scaling (cf. Remark 3.1.3).

Thus, in order to address Bohr's correspondence principle for spin systems, we must investigate the asymptotic limit and expansions, as $n \to \infty$, of the symbol correspondences $W_{\tilde{c}}^{j}$, their twisted products and the symbols themselves.

---

[1] One should also note from (3.89) and (4.11) that the $\star_{\tilde{c}}^{n}$ commutator is a derivation on the $\star_{\tilde{c}}^{n}$ algebra, while the Poisson bracket is a derivation on the algebra of functions under pointwise product.

© Springer International Publishing Switzerland 2014
P. de M. Rios, E. Straume, *Symbol Correspondences for Spin Systems*,
DOI 10.1007/978-3-319-08198-4_8

As we saw in the previous chapter, each symbol correspondence $W_{\vec{c}}^{j}$ defines a $\vec{c}$-twisted j-algebra on $Poly_{\mathbb{C}}(S^2)_{\leq n}$. However, despite the fact that all $\vec{c}$-twisted j-algebras are isomorphic for each finite j, we shall see below that only a subclass of symbol correspondence sequences yield Poisson dynamics in the asymptotic limit of high spin numbers ($j \to \infty$). This subclass realizes Rieffel's "strict deformation quantization" of the 2-sphere in reverse order (from quantum to classical). However, as we shall see below, this subclass is far from being generic.

8.1   Low-l High-j-Asymptotics of the Standard Twisted Product

If we recall from Proposition 7.1.6 the formulae for the standard twisted product of the cartesian coordinate functions, and try to compute the asymptotics for these formulae as $n \to \infty$, the first question is how to expand these formulae. A natural asymptotic expansion is in power series, but, power series of what? From Eq. (7.4), one could guess that $1/\sqrt{n(n+2)}$, or rather $1/\sqrt{j(j+1)}$, is a natural expansion parameter (as suggested in [73]). In this case, expanding in negative powers of $\sqrt{j(j+1)}$ we have

$$a \star_1^n b \longrightarrow ab + \frac{i\varepsilon_{abc}}{2\sqrt{j(j+1)}}c + O((j(j+1))^{-1}), \tag{8.1}$$

$$a \star_1^n a \longrightarrow a^2 + O((j(j+1))^{-1}). \tag{8.2}$$

On the other hand, looking at the formulae for the Berezin twisted product of the cartesian coordinate functions, as described in Corollary 7.1.11, a more natural expansion parameter seems to be $1/n$. Then, the standard twisted product of the coordinate functions in negative powers of n becomes

$$a \star_1^n b \longrightarrow ab + \frac{i\varepsilon_{abc}}{n}c + O(n^{-2}), \tag{8.3}$$

$$a \star_1^n a \longrightarrow a^2 + O(n^{-2}). \tag{8.4}$$

We also observe that in the equivalent expressions (8.1)–(8.2) and (8.3)–(8.4), the zeroth order term is the classical pointwise product, while the first-order term is the Poisson bracket (multiplied by i).

Thus we are led to inquire whether the zeroth and first-order expansions of any twisted product of any polynomial functions always coincide with the pointwise product and the Poisson bracket of these functions, respectively.

We start this investigation by studying more closely the standard Stratonovich-Weyl twisted product. But first, we have the following:

Theorem 8.1.1 ([22, 30]). *For $n = 2j \gg 1$, $l_1, l_2, l \ll 2j$,*

$$(-1)^{2j+m} \sqrt{\frac{(2j+1)(2l+1)}{(2l_1+1)(2l_2+1)}} \begin{bmatrix} l_1 & l_2 & l \\ m_1 & m_2 & -m \end{bmatrix}[j]$$

$$= C^{l_1,l_2,l}_{m_1,m_2,m} C^{l_1,l_2,l}_{0,0,0} + \frac{1}{2\sqrt{j(j+1)}} C^{l_1,l_2,l}_{m_1,m_2,m} P(l_1, l_2, l) + O((j(j+1))^{-1})$$

$$(8.5)$$

$$= C^{l_1,l_2,l}_{m_1,m_2,m} C^{l_1,l_2,l}_{0,0,0} + \frac{1}{n+1} C^{l_1,l_2,l}_{m_1,m_2,m} P(l_1, l_2, l) + O((n+1)^{-2})$$ $$(8.6)$$

$$= C^{l_1,l_2,l}_{m_1,m_2,m} C^{l_1,l_2,l}_{0,0,0} + \frac{1}{n} C^{l_1,l_2,l}_{m_1,m_2,m} P(l_1, l_2, l) + O(n^{-2})$$ $$(8.7)$$

where $C^{l_1,l_2,l}_{0,0,0}$ and $P(l_1, l_2, l)$ are given explicitly by (4.49) and (4.52), respectively, and satisfy $C^{l_1,l_2,l}_{0,0,0} \equiv 0$ if $l_1 + l_2 + l$ is odd, $P(l_1, l_2, l) \equiv 0$ if $l_1 + l_2 + l$ is even.

One can say that the zeroth order term in Eqs. (8.5)–(8.7) was first obtained by Brussard and Tolhoek in [22], though not in the form of these equations. The first order term was first obtained by Freidel and Krasnov in [30].

For the proof of Theorem 8.1.1 we refer to Appendix 10.10.

Corollary 8.1.2. *For nonnegative integers l_1, l_2, the standard twisted product of spherical harmonics $Y^{m_1}_{l_1}$ and $Y^{m_2}_{l_2}$ satisfies*

$$(i): \lim_{n\to\infty} \left(Y^{m_1}_{l_1} \star^n_1 Y^{m_2}_{l_2} - Y^{m_2}_{l_2} \star^n_1 Y^{m_1}_{l_1} \right) = 0,$$ $$(8.8)$$

$$(ii): \lim_{n\to\infty} \left(Y^{m_1}_{l_1} \star^n_1 Y^{m_2}_{l_2} + Y^{m_2}_{l_2} \star^n_1 Y^{m_1}_{l_1} \right) = 2Y^{m_1}_{l_1} Y^{m_2}_{l_2},$$ $$(8.9)$$

$$(iii): \lim_{n\to\infty} \left(n[Y^{m_1}_{l_1} \star^n_1 Y^{m_2}_{l_2} - Y^{m_2}_{l_2} \star^n_1 Y^{m_1}_{l_1}] \right) = 2i\{Y^{m_1}_{l_1}, Y^{m_2}_{l_2}\},$$ $$(8.10)$$

$$(iv): \lim_{n\to\infty} \left(n[Y^{m_1}_{l_1} \star^n_1 Y^{m_2}_{l_2} + Y^{m_2}_{l_2} \star^n_1 Y^{m_1}_{l_1} - 2Y^{m_1}_{l_1} Y^{m_2}_{l_2}] \right) = 0,$$ $$(8.11)$$

where the limits above are taken uniformly, i.e. we have uniform convergence of the sequence of functions on the l.h.s. to the function on the r.h.s.

By linearity, properties (i)–(iv) apply to the product of any $f, g \in Poly_{\mathbb{C}}(S^2)_{\leq k}$, where $k \in \mathbb{N}$ is finite.

Proof. First, for fixed l_1, m_1, l_2, m_2, let us write

$$\mathcal{P}^{l_1,l_2}_{m_1,m_2} = \{Y^m_l, |l_1 - l_2| \leq l \leq l_1 + l_2, -l \leq m = m_1 + m_2 \leq l\}$$

and start by rewriting Eq. (8.7) as

$$(-1)^{2j+m} \sqrt{\frac{(2j+1)(2l+1)}{(2l_1+1)(2l_2+1)}} \begin{bmatrix} l_1 & l_2 & l \\ m_1 & m_2 & -m \end{bmatrix}[j]$$

$$= C_{m_1,m_2,m}^{l_1,l_2,l} C_{0,0,0}^{l_1,l_2,l} + \frac{1}{n} C_{m_1,m_2,m}^{l_1,l_2,l} P(l_1,l_2,l) + \frac{1}{n^2} D_{m_1,m_2}^{l_1,l_2,l}(n) \,, \qquad (8.12)$$

where $D_{m_1,m_2}^{l_1,l_2,l}(n) \in \mathbb{R}$ is such that

$$\left| D_{m_1,m_2}^{l_1,l_2,l}(n) \right| \le K \in \mathbb{R}^+, \ \forall n \in \mathbb{N}. \qquad (8.13)$$

We will show (iii), the others follow similarly.

Let $||f|| = sup(|f(\mathbf{n})|, \mathbf{n} \in S^2)$ denote the sup-norm on the space of smooth functions on the sphere. From Theorem 8.1.1, Propositions 4.2.6 and 4.2.8 and Eq. (8.12) we have that, $\forall n \ge l_1 + l_2$,

$$\left\| n \left(Y_{l_1}^{m_1} \star_1^n Y_{l_2}^{m_2} - Y_{l_2}^{m_2} \star_1^n Y_{l_1}^{m_1} \right) - 2i\{Y_{l_1}^{m_1}, Y_{l_2}^{m_2}\} \right\| = \frac{1}{n} \|R_{l_1,l_2}^{m_1,m_2}(n)\| \,, \qquad (8.14)$$

where

$$R_{l_1,l_2}^{m_1,m_2}(n) = \sum_{l=|l_1-l_2|}^{l_1+l_2} D_{m_1,m_2}^{l_1,l_2,l}(n) Y_l^m \,, \ m = m_1 + m_2 \,.$$

Then,

$$\|R_{l_1,l_2}^{m_1,m_2}(n)\| \le \sum_{l=|l_1-l_2|}^{l_1+l_2} \|D_{m_1,m_2}^{l_1,l_2,l}(n) Y_l^m\| \le K \sum_{l=|l_1-l_2|}^{l_1+l_2} \|Y_l^m\| \,,$$

where we have used (8.13). Now, let

$$M = max\{\|Y_l^m\|, \ Y_l^m \in \mathcal{P}_{m_1,m_2}^{l_1,l_2}\} \,,$$

then

$$\|R_{l_1,l_2}^{m_1,m_2}(n)\| \le KM(l_1 + l_2 + 1 - |l_1 - l_2|) \qquad (8.15)$$

and (iii) follows immediately from (8.14) and (8.15). The other statements (i), (ii) and (iv) are proved similarly. $\square$

Remark 8.1.3. For finite $n \gg 1$, the following is a valid expansion in powers of $1/n$, as long as $l_1, l_2 \ll n$:

$$Y_{l_1}^{m_1} \star_1^n Y_{l_2}^{m_2} = Y_{l_1}^{m_1} Y_{l_2}^{m_2}|_n + \frac{i}{n}\{Y_{l_1}^{m_1}, Y_{l_2}^{m_2}\}|_n + o(1/n) \,, \qquad (8.16)$$

where $Y_{l_1}^{m_1} Y_{l_2}^{m_2}|_n$ denotes the nth degree truncation of $Y_{l_1}^{m_1} Y_{l_2}^{m_2}$ and $\{Y_{l_1}^{m_1}, Y_{l_2}^{m_2}\}|_n$ denotes the nth degree truncation of $\{Y_{l_1}^{m_1}, Y_{l_2}^{m_2}\}$, that is, the truncations $l \leq n$ of the l-summations in formulas (4.44) and (4.51), respectively.

However, the asymptotic ($n \to \infty$) expansion of the Wigner product symbol presented in Theorem 8.1.1 is invalid without the assumption $l_1, l_2 << n$ (l_1, l_2 remaining finite as $n \to \infty$) and other asymptotic expansions of the Wigner product symbol are in order if, for instance, we let $n \to \infty$ while keeping n/l_1, n/l_2, n/l finite.

8.2 Asymptotic Types of Symbol Correspondence Sequences

As the classical products of spherical harmonics appear as the zeroth and first order terms in the expansion of the standard twisted product in powers of n^{-1} (or $(j(j + 1))^{-1/2}$), we want to investigate which of all possible symbol correspondences have the same property, namely that their twisted products are related to the classical products of functions on the sphere, as $n \to \infty$. By linearity, it is enough to investigate this for the products of spherical harmonics.

However, again turning to the standard Berezin twisted product of linear spherical harmonics or cartesian coordinate functions (cf. Corollary 7.1.11), we note that the zeroth order term in the $1/n$ (or $1/\sqrt{j(j+1)}$) expansion coincides with the pointwise product, whereas the first-order term does not coincide with the Poisson bracket of functions. Therefore, first we introduce the following:

Definition 8.2.1. Let

$$\Delta^+(\mathbb{N}^2) = \{(n, l) \in \mathbb{N}^2 \mid n \geq l > 0\}$$

and $\mathcal{C} : \Delta^+(\mathbb{N}^2) \to \mathbb{R}^*$ be any given function. We denote by $\mathbf{W}_{\mathcal{C}} = [W_{\tilde{c}}^j]_{2j=n\in\mathbb{N}}$ the *sequence of symbol correspondences* defined by characteristic numbers $c_l^n = \mathcal{C}(n, l), \forall (n, l) \in \Delta^+(\mathbb{N}^2), c_0^n = 1, \forall n \in \mathbb{N}$. We denote by

$$\mathbf{W}_{\mathcal{C}}(S^2, \star) = [(Poly_{\mathbb{C}}(S^2)_{\leq n}, \star_{\tilde{c}}^n)]_{n\in\mathbb{N}}$$

the associated *sequence of twisted algebras* (cf. Definitions 6.2.22 and 7.1.16).

8.2.1 Symbol Correspondence Sequences of Poisson Type

Definition 8.2.2. A symbol correspondence sequence $\mathbf{W}_{\mathcal{C}}$, with its associated sequence of twisted algebras $\mathbf{W}_{\mathcal{C}}(S^2, \star)$, is of *Poisson type* if, $\forall l_1, l_2 \in \mathbb{N}$,

$$(\text{i}): \lim_{n \to \infty} \left(Y_{l_1}^{m_1} \star_{\tilde{c}}^n Y_{l_2}^{m_2} - Y_{l_2}^{m_2} \star_{\tilde{c}}^n Y_{l_1}^{m_1} \right) = 0, \tag{8.17}$$

$$(\text{ii}): \lim_{n \to \infty} \left(Y_{l_1}^{m_1} \star_{\tilde{c}}^n Y_{l_2}^{m_2} + Y_{l_2}^{m_2} \star_{\tilde{c}}^n Y_{l_1}^{m_1} \right) = 2 Y_{l_1}^{m_1} Y_{l_2}^{m_2}, \tag{8.18}$$

$$(\text{iii}): \lim_{n \to \infty} \left(n[Y_{l_1}^{m_1} \star_{\tilde{c}}^n Y_{l_2}^{m_2} - Y_{l_2}^{m_2} \star_{\tilde{c}}^n Y_{l_1}^{m_1}] \right) = 2i \{ Y_{l_1}^{m_1}, Y_{l_2}^{m_2} \}, \tag{8.19}$$

$\mathbf{W}_C$ is of *anti-Poisson type* if it satisfies properties (i), (ii) above and

$$(\text{iii'}): \lim_{n \to \infty} \left(n[Y_{l_1}^{m_1} \star_{\tilde{c}}^n Y_{l_2}^{m_2} - Y_{l_2}^{m_2} \star_{\tilde{c}}^n Y_{l_1}^{m_1}] \right) = -2i \{ Y_{l_1}^{m_1}, Y_{l_2}^{m_2} \}, \tag{8.20}$$

and $\mathbf{W}_C$ is of *pure-*(resp. *pure-anti*)*-Poisson type* if, in addition to properties (i), (ii), and (iii) (resp. (iii')) above, the following property also holds:

$$(\text{iv}): \lim_{n \to \infty} \left(n[Y_{l_1}^{m_1} \star_{\tilde{c}}^n Y_{l_2}^{m_2} + Y_{l_2}^{m_2} \star_{\tilde{c}}^n Y_{l_1}^{m_1} - 2Y_{l_1}^{m_1} Y_{l_2}^{m_2}] \right) = 0. \tag{8.21}$$

Again, all limits above are taken uniformly, i.e. we consider uniform convergence of the sequence of functions on the l.h.s. to the function on the r.h.s.

Remark 8.2.3. The signs in the r.h.s. of Eqs. (8.19) and (8.20) are related to the choice of orientation of the symplectic form on the sphere (cf. Remark 4.1.3). Once a choice is fixed, they are also related by the antipodal map on the sphere (cf. Proposition 6.2.36, Eq. (6.41), and Proposition 7.1.13).

Remark 8.2.4. If $\mathbf{W}_C$ is of pure-Poisson type, its twisted product expands as in Eq. (8.16), under the same assumptions of Remark 8.1.3.

Proposition 8.2.5. *The standard Stratonovich-Weyl symbol correspondence sequence is of pure-Poisson type and the alternate Stratonovich-Weyl symbol correspondence sequence is of pure-anti-Poisson type.*

Proof. For the standard case, this follows from Corollary 8.1.2. For the alternate case, $c_l^n = \varepsilon_l = (-1)^l$, it follows from the above and from Proposition 7.1.13.   □

Proposition 8.2.6. *The standard (resp. alternate) Berezin symbol correspondence sequence is of Poisson type (resp. anti-Poisson type), but not of pure-Poisson type (resp. pure-anti-Poisson type). Same for the standard (resp. alternate) Toeplitz symbol correspondence sequences (cf. Definition 7.2.27).*

Proof. To see that the standard Berezin twisted product is of Poisson type, note from formula (6.38) for $b_l^n = \mathcal{C}(n, l)$ that

$$\forall l, l_1, l_2 << n, \quad \frac{b_l^n}{b_{l_1}^n b_{l_2}^n} \longrightarrow 1 + O(1/n), \text{ as } n \to \infty.$$

Therefore, by (7.9) and (8.7), $\forall l_1, l_2 << n,$

$$\lim_{n\to\infty} \left(Y_{l_1}^{m_1} \star_{\tilde{b}}^{n} Y_{l_2}^{m_2} \right) = \lim_{n\to\infty} \left(Y_{l_1}^{m_1} \star_{1}^{n} Y_{l_2}^{m_2} \right),$$

$$\lim_{n\to\infty} \left(n[Y_{l_1}^{m_1} \star_{\tilde{b}}^{n} Y_{l_2}^{m_2} - Y_{l_2}^{m_2} \star_{\tilde{b}}^{n} Y_{l_1}^{m_1}] \right) = \lim_{n\to\infty} \left(n[Y_{l_1}^{m_1} \star_{1}^{n} Y_{l_2}^{m_2} - Y_{l_2}^{m_2} \star_{1}^{n} Y_{l_1}^{m_1}] \right).$$

On the other hand, property (iv) already fails for the twisted product of linear symbols, as shown by Corollary 7.1.11. For the alternate case, $\mathcal{C}(n,l) = b_{l_-}^n = (-1)^l b_l^n$, we use Proposition 7.1.13. Analogously for the standard and alternate Toeplitz twisted products, cf. Definition 7.2.25 and Eq. (7.82). □

Remark 8.2.7. The distinction between symbol correspondence sequences of pure-Poisson or Poisson types is not irrelevant insofar as the Stratonovich-Weyl and the Berezin and Toeplitz symbol correspondences satisfy different axioms, namely, the former satisfies the isometry axiom (v) of Remark 6.1.2 while the latter ones do not. It is therefore interesting to see that these distinct symbol correspondence sequences exhibit distinct asymptotics, namely the former satisfies the property (8.21) while the latter ones do not.

We shall say more about their asymptotics below, when a more important asymptotic distinction between the standard Stratonovich-Weyl and the Berezin and Toeplitz correspondences will be highlighted (see Remarks 8.2.26 and 8.2.28).

8.2.2 Symbol Correspondence Sequences of Non-Poisson Type

Generic symbol correspondence sequences are not of Poisson or anti-Poisson type. This can already be seen in the very restrictive case of Stratonovich-Weyl symbol correspondences. Thus, consider a Stratonovich-Weyl correspondence sequence given by the characteristic numbers $c_l^n = \varepsilon_l^n = \pm 1$, for $1 \le l \le n$. Combining Eqs. (7.9) and (8.7), we have for $l_1, l_2 \ll n$,

$$Y_{l_1}^{m_1} \star_{\varepsilon}^{n} Y_{l_2}^{m_2} = \sum_{\substack{l=|l_1-l_2| \\ l \equiv l_1+l_2 (\text{mod } 2)}}^{l_1+l_2} \sqrt{\frac{(2l_1+1)(2l_2+1)}{2l+1}} \, C_{m_1,m_2,m}^{l_1,l_2,l} C_{0,0,0}^{l_1,l_2,l} \frac{\varepsilon_l^\infty}{\varepsilon_{l_1}^\infty \varepsilon_{l_2}^\infty} Y_l^m$$

(8.22)

$$+ \frac{1}{n} \sum_{\substack{l=|l_1-l_2|+1 \\ l \equiv l_1+l_2-1 (\text{mod } 2)}}^{l_1+l_2-1} \sqrt{\frac{(2l_1+1)(2l_2+1)}{2l+1}} \, C_{m_1,m_2,m}^{l_1,l_2,l} P(l_1,l_2,l) \frac{\varepsilon_l^\infty}{\varepsilon_{l_1}^\infty \varepsilon_{l_2}^\infty} Y_l^m + O(1/n^2)$$

where

$$\varepsilon_l^\infty = \lim_{n\to\infty} \varepsilon_l^n,$$

(8.23)

whenever such limits exist.

Clearly, limit (8.23) does not exist for a random sequence of strings of ± 1, of the form $\vec{\varepsilon} = (\varepsilon_1^n, \varepsilon_2^n, \cdots, \varepsilon_n^n)$. Thus, obviously, these generic Stratonovich-Weyl symbol correspondences are not of Poisson type.

Moreover, by comparison with Eqs. (4.44) and (4.51), we see that the same can be said of a generic string of ± 1 of the form $\vec{\varepsilon} = (\varepsilon_1, \varepsilon_2, \cdots, \varepsilon_n)$, where $\varepsilon_l^n = \varepsilon_l = \varepsilon_l^\infty$ because, generically, $\varepsilon_l / \varepsilon_{l_1} \varepsilon_{l_2}$ will be a random assignment

$$\epsilon_{l_1, l_2} : \mathbb{N} \cap [|l_1 - l_2| - 1, l_1 + l_2] \to \{\pm 1\}$$

and thus, generically, the first sum in (8.22) will not yield the pointwise product $Y_{l_1}^{m_1} Y_{l_2}^{m_2}$ and the second sum will not yield the Poisson bracket (times $\pm i$) $\{Y_{l_1}^{m_1}, Y_{l_2}^{m_2}\}$.

In fact, the requirement that, for every $Y_{l_1}^{m_1}$ and $Y_{l_2}^{m_2}$, $l_1, l_2 << n$, the first sum in (8.22) yields the pointwise product $Y_{l_1}^{m_1} Y_{l_2}^{m_2}$ and the second sum yields the Poisson bracket (times $\pm i$) $\{Y_{l_1}^{m_1}, Y_{l_2}^{m_2}\}$, enforces that, $\forall l, l_1, l_2 << n$, either $\varepsilon_l / \varepsilon_{l_1} \varepsilon_{l_2} = 1$ or $\varepsilon_l / \varepsilon_{l_1} \varepsilon_{l_2} = (-1)^{l + l_1 + l_2}$ (cf. Proposition 7.1.13). Thus, we have:

Proposition 8.2.8. *A Stratonovich-Weyl symbol correspondence sequence* $[\varepsilon_l^n]$ *which is of Poisson type is a sequence for which* $\varepsilon_l^\infty = 1$, $\forall l \in \mathbb{N}$, *and it is of anti-Poisson type if* $\varepsilon_l^\infty = (-1)^l$, $\forall l \in \mathbb{N}$ *(cf. Eq. (8.23)).*

Although a generic symbol correspondence sequence has no limit as $n \to \infty$, there exist symbol correspondence sequences which are not of Poisson or anti-Poisson type, but still have a well-defined limit as $n \to \infty$. This is interesting because such a correspondence sequence defines a dynamics of symbols which is not of Poisson type, in the limit $n \to \infty$. We illustrate this with some examples:

Example 8.2.9. Let $\mathbf{W}_{\mathcal{C}}$ be the symbol correspondence sequence defined by the characteristic numbers $c_l^n = \mathcal{C}(n, l) = n^{-l}$. As $n \to \infty$, the twisted products of the cartesian symbols expand as

$$x \star_{\hat{c}}^n y = xy + iz + O(1/n), \quad x \star_{\hat{c}}^n x = x^2 + 1/2 + O(1/n) \qquad (8.24)$$

and so on for cyclic permutations of (x, y, z). We note that the commutator

$$x \star_{\hat{c}}^n y - y \star_{\hat{c}}^n x = 2iz + O(1/n) = 2i\{x, y\} + O(1/n),$$

so the Poisson bracket appears as the zeroth order term in the expansion of the commutator, not as the first order term, as in Eq. (8.19). Thus, this symbol correspondence sequence is not of Poisson type, according to Definition 8.2.2.

For this symbol correspondence sequence, however, one could try redefining Poisson dynamics as the zeroth order term in the expansion of the commutator. But this clearly does not work either, because

$$\frac{c_l^n}{c_{l_1}^n c_{l_2}^n} = n^{l_1 + l_2 - l}$$

and therefore, using Eq. (7.9), we can see that the expansion in powers of n, or $1/n$, of the twisted product $Y_{l_1}^{m_1} \star_{\tilde{c}}^n Y_{l_2}^{m_2}$ will be completely messed up, with each expanding term power depending on $l_1 + l_2 - l$, for $|l_1 - l_2| \leq l \leq l_1 + l_2$.

Example 8.2.10. Let $\mathbf{W}_C$ be the symbol correspondence sequence defined by the characteristic numbers $c_l^n = C(n, l) = l b_l^n - (l - 1)/n^l$, where b_l^n are the characteristic numbers of the standard Berezin symbol correspondence, given by (6.37). Then, clearly,

$$\forall l, l_1, l_2 << n, \ \frac{c_l^n}{c_{l_1}^n c_{l_2}^n} \to \frac{l}{l_1 l_2} + O(1/n), \ as \ n \to \infty.$$

Inserting the above estimate in Eq. (7.9), using Eq. (8.7), we have

$$
\lim_{n \to \infty} Y_{l_1}^{m_1} \star_{\tilde{c}}^n Y_{l_2}^{m_2}
$$

$$
= \sum_{\substack{l = |l_1 - l_2| \\ l \equiv l_1 + l_2 (\mathrm{mod}\ 2)}}^{l_1 + l_2} \sqrt{\frac{(2l_1 + 1)(2l_2 + 1)}{2l + 1}} \ C_{m_1, m_2, m}^{l_1, l_2, l} C_{0,0,0}^{l_1, l_2, l} \frac{l}{l_1 l_2} Y_l^m \tag{8.25}
$$

$$
\lim_{n \to \infty} \left(n [Y_{l_1}^{m_1} \star_{\tilde{c}}^n Y_{l_2}^{m_2} - Y_{l_2}^{m_2} \star_{\tilde{c}}^n Y_{l_1}^{m_1}] \right)
$$

$$
= \sum_{\substack{l = |l_1 - l_2| + 1 \\ l \equiv l_1 + l_2 - 1 (\mathrm{mod}\ 2)}}^{l_1 + l_2 - 1} \sqrt{\frac{(2l_1 + 1)(2l_2 + 1)}{2l + 1}} \ C_{m_1, m_2, m}^{l_1, l_2, l} P(l_1, l_2, l) \frac{l}{l_1 l_2} Y_l^m \tag{8.26}
$$

and from (4.44) and (4.51) we immediately see that the first-order expansion in $1/n$ of the commutator of $\star_{\tilde{c}}^n$ does not coincide with the Poisson bracket, nor does the zeroth order expansion in $1/n$ of $\star_{\tilde{c}}^n$ coincide with the pointwise product. Therefore, this symbol correspondence sequence is not of Poisson type, either.

Now, one might think that these two examples above seem ad-hoc and that more "realistic" symbol correspondences would not exhibit such "exotic" behaviors. However, the following example, the example of the upper-middle-state correspondence (cf. Definition 6.2.55), is enough to dissipate this prejudice.

We recall that the upper-middle-state correspondence is a mapping-positive symbol correspondence defined via (6.13) by an operator kernel K which is a J_3-invariant state, so that K^g is a coherent (family of) state(s).

Thus, the upper-middle-state correspondence and its antipodal, the lower-middle-state correspondence, belong to the set of "most realistic possible" symbol correspondences, which are the ones obtained by a coherent (family of) state(s), realizing quantum expectation values. Furthermore, this particular coherent state is defined in a simple invariant and n-independent way via Proposition 6.2.59.

Example 8.2.11. Let $\mathbf{W}_C$ be the upper-middle-state symbol correspondence sequence, which is defined by the characteristic numbers (6.52)–(6.55), or equivalently, by Proposition 6.2.59. Then, the characteristic numbers of this symbol correspondence satisfy

$$\lim_{n\to\infty} p_l^n(1/2) = \frac{(-1)^{l/2}}{2^l}\binom{l}{l/2}, \text{ when } l \text{ is even}, \qquad (8.27)$$

$$= 0 \ (+\ O(1/n)), \text{ when } l \text{ is odd}, \qquad (8.28)$$

cf. (6.57), (6.52)–(6.55), and we can easily see that the upper-middle-state symbol correspondence sequence is very far from being of Poisson type. Same for the lower-middle-state correspondence, which stands in alternate relation to this one.

Let us expand on this example. First, from (7.1.8), (8.1.1) and Proposition 7.1.15, we see that the expansion of the twisted anti-commutator $[[\ ,\ ]]_\star$ and twisted commutator $[\ ,\ ]_\star$ have the asymptotic orders as outlined below:

$$[[\ even, even\ ]]_\star = O(1)\{even\} \qquad [\ odd, odd\ ]_\star = O(1)\{odd\} \qquad (8.29)$$

$$[[\ odd, odd\ ]]_\star = ``O(n^2)"\{even\} \qquad [\ even, even\ ]_\star = O(1/n^2)\{odd\} \qquad (8.30)$$

$$[[\ even, odd\ ]]_\star = O(1)\{odd\} \qquad [\ odd, even\ ]_\star = O(1)\{even\} \qquad (8.31)$$

where, following notation of Proposition 7.1.15, $[[\ even, even\ ]]_\star = O(1)\{even\}$ means that the anti-commutator of two even functions is an even function, in order zero in the expansion in powers of $1/n$, and similarly, $[\ even, even\ ]_\star = O(1/n^2)\{odd\}$ means that the commutator of two even functions is an odd function, in order $1/n^2$, but vanishes in $O(1)$ and $O(1/n)$, etc., *assuming* that all *even* and *odd* functions appearing in the l.h.s. and r.h.s. of these identities are $O(1)$.

Note also from (8.27) that the zeroth order expansion of the even characteristic numbers depends strongly on l (likewise, one obtains from (6.53)–(6.55) that the first order expansion for the odd characteristic numbers also depends strongly on l), so that taking the anti-commutator of two even functions, for instance, will not generally give (twice) the pointwise product of these functions in zeroth order.

Also, the leftmost identity in (8.30) indicates that the anti-commutator of two odd functions seems to "blow up" quadratically with n, as $n \to \infty$, which at first makes no sense within an asymptotic expansion of the twisted product in positive powers of $1/n$. Thus, we will now provide an interpretation for this term.

To approach an interpretation for this "blow-up", let us start by focusing on the correspondence induced on the $l = 1$ subspace, with its first characteristic number c_1^n. Note from Eqs. (6.53)–(6.55) that we have

$$p_1^n(1/2) = c_1^n = 1/\sqrt{n(n+2)} = 1/n + O(1/n^2) \, , \tag{8.32}$$

so that the symbol correspondence in the $l = 1$ subspace becomes

$$\frac{J_1}{\sqrt{j(j+1)}} \mapsto c_1^n x \, , \quad \frac{J_2}{\sqrt{j(j+1)}} \mapsto c_1^n y \, , \quad \frac{J_3}{\sqrt{j(j+1)}} \mapsto c_1^n z \tag{8.33}$$

(each J_i a $(n+1)$-square matrix), with c_1^n given by (8.32), cf. Remark 7.1.7.

At first, one could think that for this correspondence, the image of the Lie algebra of $SU(2)$ in the space of polynomials, which as a subalgebra of the Poisson-Lie algebra is generated by the linear monomials x, y, z, would vanish in the asymptotic limit $n \to \infty$ (compare with (7.8)). But this is an oversimplification.

In fact we note, upon closer analysis, that the correspondence given by (8.33) has non-zero right-hand sides for every finite $n \in \mathbb{N}$ and the limit $n \to \infty$ is an asymptotic limit, so that $n = \infty$ is never actually realized (see further discussion on this point in Sect. 8.3). And as we discuss below, the non-vanishing of these linear symbols is not only relevant, but also has nontrivial consequences.

Furthermore, the rightmost identities in (8.29) and (8.31) are explicit indications that odd symbols (linear ones included) generate proper dynamics on the whole space of symbols, in the asymptotic limit (although this asymptotic dynamics will not generally be Poissonian), and therefore cannot be neglected.

However, we do have the interesting fact that, under the upper-middle-state correspondence, the linear symbols x, y, z become *subdominant* in the asymptotic limit, in comparison with the even symbols, like x^2, y^2, yz, zx, etc., which are of order zero in $1/n$. And in view of (8.27)–(8.28), this subdominance extends to all symbols that are odd polynomials, in comparison with the even polynomials.

In this light, we reinterpret the leftmost identity in (8.30) as indicating an asymptotic phenomenon, which is a kind of asymptotic "algebraic resurgence", when two subdominant symbols multiply (anticommute) to become dominant.

So, let us again take a closer look at linear symbols and their twisted products, under the upper-middle-state correspondence. In order to obtain more precise formulae for these products, we will follow the same method used for the computation of the standard Berezin twisted product of linear symbols, outlined before Corollary 7.1.11. Thus, we need more precise formulae for the two first characteristic numbers. The first one is given by (8.32). For the second, we have:

$$p_2^n(1/2) = c_2^n = -\frac{1}{2}\pi_n \, , \quad \text{when } n \text{ is odd} \, , \tag{8.34}$$

$$= -\frac{1}{2}\pi_n \left(1 + O(1/n^2)\right) \, , \quad \text{when } n \text{ is even} \, , \tag{8.35}$$

where π_n is given in (7.7), that is

$$\pi_n = \sqrt{\frac{(n-1)(n+3)}{n(n+2)}}.$$ (8.36)

Notation 8.2.12. *Let us now set the following notation:*

$$f \approx g \iff f = g(1 + O(1/n^2)).$$ (8.37)

Then, following the same method used to obtain Corollary 7.1.11, we obtain:

Proposition 8.2.13. *The upper-middle-state twisted product $\star_{\frac{n}{p}}$ of the cartesian coordinate functions $\{a, b, c\} = \{x, y, z\}$ are given by*

$$a \star_{\frac{n}{p}} b \approx -\frac{(n-1)(n+3)}{2}ab + i\varepsilon_{abc}c,$$ (8.38)

$$a \star_{\frac{n}{p}} a \approx -\frac{(n-1)(n+3)}{2}a^2 + \frac{(n+1)^2}{2} - 1,$$ (8.39)

$$a \star_{\frac{n}{p}} a + b \star_{\frac{n}{p}} b + c \star_{\frac{n}{p}} c \approx n(n+2).$$ (8.40)

Remark 8.2.14. The $\approx$ sign in (8.38)–(8.40) is actually an equality ($=$) when n is odd, in view of (8.34).

Therefore, by taking anti-commutators and commutators of the products above, we see them as special cases of the leftmost identity in (8.30) and the rightmost identity in (8.29). Specifically, the commutator is (the $\mathcal{SU}(2)$ algebra)

$$[x, y]_{\star_{\frac{n}{p}}} = 2iz,$$ (8.41)

plus cyclic permutations. Now identifying $W_x^j = c_1^n x$, $W_y^j = c_1^n y$, $W_z^j = c_1^n z$, from (8.33), then the Eq. (8.41) for the Lie algebra of $SU(2)$ becomes

$$[W_x^j, W_y^j]_{\star_{\frac{n}{p}}} = (c_1^n)^2 2iz = c_1^n 2i W_z^j = \frac{2i W_z^j}{\sqrt{n(n+2)}},$$ (8.42)

which has the same form of the commutator obtained from the standard twisted product, cf. (7.4), as one should expect, because the Lie algebra of linear symbols (multiplied by the first characteristic number) equals the Lie algebra of operators.

However, for the purpose of our discussion regarding asymptotic orders of the products, the relevant fact is that (8.42) has the regular asymptotic order. In the same vein, taking the anti-commutator of the regular linear symbols, we get

$$[[W_a^j, W_b^j]]_{\star_{\frac{n}{p}}} \approx -(\pi_n)^2 ab + \frac{n^2 + 2n - 1}{n^2 + 2n}\delta_{a,b}$$ (8.43)

where the sign $\approx$ is an equality when n is odd and $\delta_{a,b}$ is the Kronecker delta.

Therefore, Eq. (8.43) provides an explicit instance of the asymptotic algebraic resurgence mentioned before, as the symbols anti-commuting on the l.h.s. are each of order $O(1/n)$, while the symbol on the r.h.s. is of order $O(1)$.

Thus, if we denote by $\widetilde{odd}$ the odd symbols with their correct asymptotic order, which is $O(1/n)$ according to (8.28), then we can rewrite (8.29)–(8.31) as

$$[[\,even, even\,]]_\star = O(1)\{even\} \qquad [\,\widetilde{odd}, \widetilde{odd}\,]_\star = O(1/n)\{\widetilde{odd}\} \qquad (8.44)$$

$$[[\,\widetilde{odd}, \widetilde{odd}\,]]_\star = O(1)\{even\} \qquad [\,even, even\,]_\star = O(1/n)\{\widetilde{odd}\} \qquad (8.45)$$

$$[[\,even, \widetilde{odd}\,]]_\star = O(1)\{\widetilde{odd}\} \qquad [\,\widetilde{odd}, even\,]_\star = O(1/n)\{even\} \qquad (8.46)$$

where in the r.h.s. of the rightmost identity in (8.44), for instance, $O(1/n)\{\widetilde{odd}\}$ means that this term is of order $O(1/n)$ *with respect to* the order of $\widetilde{odd}$, and similarly for all the other cases. Therefore, now all identities in (8.44)–(8.46) have the regular asymptotic orders and nothing blows up.

We emphasize, however, that having regularized the asymptotic orders does not mean, at all, that the twisted products have become Poissonian in the asymptotic limit. This was already made clear by considering the leftmost identity in (8.44), in view of (8.27), and is clearer when considering the leftmost identity in (8.45), which for just checking Poisson vs. non-Poisson, is equivalent to (8.30).

8.2.3 Refined Types of Symbol Correspondence Sequences

In the last section, we presented various cases of symbol correspondences of non-Poisson type and, in particular, considered with some care three examples: Examples 8.2.9 and 8.2.10, as well as the upper-middle-state symbol correspondence, Example 8.2.11. However, although all these three examples exhibit non-Poissonian dynamics of symbols in the asymptotic limit $n \to \infty$, their symbol correspondence sequences should not be regarded as having the same asymptotic type.

Namely, the characteristic numbers in Example 8.2.9 satisfy

$$\lim_{n\to\infty} c_l^n = 0\,, \ \forall l \in \mathbb{N},$$

whereas the characteristic numbers in Example 8.2.10 satisfy

$$0 < \lim_{n\to\infty} c_l^n = c_l^\infty < \infty\,, \ \forall l \in \mathbb{N},$$

and the characteristic numbers of the upper-middle-state correspondence satisfy

$$-\infty < \lim_{n\to\infty} c_l^n = c_l^\infty < \infty\,, \ \forall l = 2q,\ q \in \mathbb{N},$$

$$\lim_{n\to\infty} c_l^n = 0\,, \ \forall l = 2q+1,\ q \in \mathbb{N}\cup\{0\}.$$

Furthermore, we saw at the beginning of the previous section that generic symbol correspondence sequences may not have a well-defined limit. In view of all this, it is necessary to refine our distinction of symbol correspondence sequences.

Definition 8.2.15. Let $\mathcal{C} : \Delta^+(\mathbb{N}^2) \to \mathbb{R}^*$, cf. Definition 8.2.1. The symbol correspondence sequence $\mathbf{W}_\mathcal{C}$ determined by characteristic numbers $c_0^n = 1, \forall n \in \mathbb{N}$, $c_l^n = \mathcal{C}(n, l)$, $\forall (n, l) \in \Delta^+(\mathbb{N}^2)$, is of *limiting type* if

$$\exists \lim_{n \to \infty} c_l^n = c_l^\infty , \ \forall l \in \mathbb{N}, \tag{8.47}$$

and it is of *strong-limiting type* if, in addition,

$$\exists \lim_{(n,l)_{l \le n} \to (\infty,\infty)} |c_l^n| . \tag{8.48}$$

Remark 8.2.16. In the above definition, the limits are taken in the usual way. Thus, (8.47) means that, $\forall l \in \mathbb{N}, \exists \lambda(l) \in \mathbb{R}$, s.t. $\forall \epsilon > 0, \exists k(l, \epsilon) \in \mathbb{N}$ s.t. $n > k \Rightarrow |c_l^n - \lambda(l)| < \epsilon$. And (8.48) means that, $\exists \eta \in \mathbb{R}$ s.t. $\forall \delta > 0, \exists p(\delta), q(\delta) \in \mathbb{N}$ s.t. $n \ge l, n > p, l > q \Rightarrow ||c_l^n| - \eta| < \delta$.

And as we saw before, among symbol correspondence sequences of limiting type there can be quite distinct limiting behaviors. It is clear from the example of the upper-middle-state correspondence, as well as Example 8.2.9, that having some characteristic numbers converging to zero in the asymptotic limit leads to particularly strong non-classical behavior/dynamics for the symbols.

Therefore, we would like to distinguish some "more tame" types of symbol correspondence sequences of non-Poisson type. We present the following definition:

Definition 8.2.17. A limiting-type symbol correspondence sequence is of *pseudo-classical type* if

$$0 < \lim_{n \to \infty} |c_l^n| < \infty , \forall l \in \mathbb{N}, \tag{8.49}$$

and it is of *quasi-classical type* if

$$\lim_{n \to \infty} |c_l^n| = 1 , \forall l \in \mathbb{N}, \tag{8.50}$$

while a symbol correspondence sequence that satisfies (8.50) but which is not necessarily of limiting type shall be called a symbol correspondence sequence of *ASD type* (cf. Remark 8.2.20, below).

The symbol correspondence sequence in Example 8.2.9 is of strong-limiting type, but not of pseudo-classical type, while the one in Example 8.2.10 is of pseudo-classical type, but not of strong-limiting type. On the other hand, the upper-middle and lower-middle-state correspondence sequences are of limiting type, but they are neither of strong-limiting type nor of pseudo-classical type.

Proposition 8.2.18. *Both the standard and the alternate Stratonovich-Weyl symbol correspondence sequences are of quasi-classical type and of strong-limiting type. On the other hand, both the standard and the alternate Berezin and Toeplitz symbol correspondence sequences are of quasi-classical type, but not of strong-limiting type.*

Proof. The first statement is obvious. The quasi-classical property in the Berezin and Toeplitz cases follow from the expansion (6.38) for b_l^n, that is,

$$\lim_{n \to \infty} b_l^n = b_l^\infty = 1 \,, \forall l \in \mathbb{N}. \tag{8.51}$$

On the other hand, note from the closed formula (6.37) for b_l^n that

$$\lim_{n \to \infty} b_n^n = 0, \tag{8.52}$$

and thus, Eqs. (8.51)–(8.52) imply that (8.48) is not satisfied for the standard and alternate Berezin and Toeplitz symbol correspondence sequences. □

Example 8.2.19. The symbol correspondence sequence with $c_0^n = 1$, $c_l^n = C(n, l) = 1 - \log(1 - ((l - 1)/n))$, $l > 0$, and the one with $C(n, l) = 1 + n^{-(1/l)}$, are both of quasi-classical type, but not of strong-limiting type.

Note that every Stratonovich-Weyl symbol correspondence sequence is of ASD type, regardless of whether it is of Poisson type or of quasi-classical type.

Remark 8.2.20. (i) In view of Theorem 6.2.21 and Remark 6.2.25, we can say that a symbol correspondence sequence of ASD type is *asymptotically self-dual*, at least for finite l's (we recall that the set of quasi-classical correspondences is a proper subset of the set of ASD correspondences, cf. Definition 8.2.17).

In view of Eq. (6.25), this self-duality is an important (classical) feature for a symbol correspondence sequence. Thus, Eq. (8.52) means that the standard and alternate Berezin and Toeplitz symbol correspondence sequences loose this important asymptotic self-dual property if $l \to \infty$ as $n \to \infty$.

(ii) On the other hand, from Eqs. (8.51), (7.45) and (7.90) it follows that the Berezin transform, given by Eq. (7.92), tends, as $n \to \infty$, to the identity on $Poly_{\mathbb{C}}(S^2)_{\leq n}$, when applied to functions which are decomposable in a finite sum of spherical harmonics (finite l's). The same can be said, of course, of its unique positive square root, the Stratonovich-Berezin transform, given by Eq. (7.96), and their respective inverses, given by Eqs. (7.93) and (7.95).

Now, the standard and alternate Stratonovich-Weyl, as well as standard and alternate Berezin correspondences, which are of Poisson-type, are also of quasi-classical type. However, it is important to know if/when the converse is true.

Furthermore, as we have seen from Examples 8.2.9 and 8.2.10 and from the examples of the upper and lower-middle-state correspondences, there are various kinds of symbol correspondence sequences of limiting type that define asymptotic

$(n \to \infty)$ dynamics of symbols which does not coincide with Poisson dynamics, and therefore it is important to know exactly which ones are of Poisson type.

But in fact, the following are necessary and sufficient conditions for a symbol correspondence sequence to be of Poisson, or pure-Poisson type:

Theorem 8.2.21. *A symbol correspondence sequence* $\mathbf{W}_C$ *is of Poisson type if and only if its characteristic numbers* $c_l^n = C(n, l)$ *satisfy*

$$\lim_{n \to \infty} C(n, l) = c_l^{\infty} = 1, \tag{8.53}$$

$\mathbf{W}_C$ *is of anti-Poisson type if and only if*

$$\lim_{n \to \infty} C(n, l) = c_l^{\infty} = (-1)^l, \tag{8.54}$$

and $\mathbf{W}_C$ *is of pure-(resp. anti)-Poisson type if*

$$\lim_{n \to \infty} n(C(n, l) - c_l^{\infty}) = 0, \tag{8.55}$$

where $c_l^{\infty} = 1$ *(resp.* $c_l^{\infty} = (-1)^l$*),* $\forall l \in \mathbb{N}$.

Proof. For the $c_l^{\infty} \equiv 1$ case, sufficiency of condition (8.55) follows from Eqs. (7.9) and (8.7) and Corollary 8.1.2. The weaker condition (8.53) uses the fact that the first term of the expansion (8.7) is commutative, while the second is anti-commutative. The case $c_l^{\infty} = (-1)^l$ follows from Proposition 7.1.13.

On the other hand, from Eqs. (3.110), (7.9), (8.7), and (8.17)–(8.19), it follows as in Proposition 8.2.8 that $\mathbf{W}_C$ is of Poisson type, or anti-Poisson type only if, $\forall l, l_1, l_2 \in \mathbb{N}$, either $\lim_{n \to \infty} \dfrac{c_l^n}{c_{l_1}^n c_{l_2}^n} = 1$, or $\lim_{n \to \infty} \dfrac{c_l^n}{c_{l_1}^n c_{l_2}^n} = (-1)^{l_1 + l_2 + l}$, implying condition (8.53); that is, either $\lim_{n \to \infty} C(n, l) = c_l^{\infty} \equiv 1$, or $\lim_{n \to \infty} C(n, l) = c_l^{\infty} = (-1)^l$. Necessity of (8.55) for the pure-Poisson case follows analogously. $\square$

Corollary 8.2.22. *Any symbol correspondence sequence of Poisson type is of quasi-classical type, but the converse is generically not true for sequences which are not characteristic-positive (or characteristic-alternate, for the anti-Poisson case).*

Example 8.2.23. A symbol correspondence sequence such that

$$\lim_{n \to \infty} C(n, l) = c_l^{\infty} = \begin{cases} -1 & \text{if } l \equiv 1 \bmod 3 \\ 1 & \text{otherwise} \end{cases}$$

is a symbol correspondence sequence of quasi-classical type which is not of Poisson nor of anti-Poisson type (same, of course, if we change mod 3 to mod 4, 5, etc.).

Remark 8.2.24. However, we note that any symbol correspondence sequence of pseudo-classical type can in principle be "renormalized" to become a symbol correspondence sequence of Poisson type, if the limit of every sequence of characteristic numbers is known. In fact, if $c_l^n = C(n, l)$ denote the characteristic numbers of a symbol correspondence sequence of pseudo-classical type, then the "renormalized" symbol correspondence sequence with characteristic numbers $\chi_l^n = c_l^n / c_l^\infty$ is of Poisson type, where $0 \neq c_l^\infty = \lim_{n \to \infty} c_l^n$.

Finally, even among the symbol correspondence sequences of Poisson type, we would like to establish some distinctions. One such has already been in use: Poisson versus pure-Poisson type, distinguishing the standard Berezin and standard Stratonovich-Weyl correspondences, for instance. However, by comparing these two we note a further distinction, leading to our last definition, which apparently expresses the optimal asymptotic conditions for symbol correspondence sequences.

Definition 8.2.25. A symbol correspondence sequence of Poisson (resp. anti-Poisson) type which is also of strong-limiting type is of *Bohr* (resp. *anti-Bohr*) type. A symbol correspondence sequence of pure- (resp. pure-anti)-Poisson type which is also of strong-limiting type is of *pure-*(resp. *pure-anti*)-*Bohr type*.

Remark 8.2.26. The standard (resp. alternate) Berezin and Toeplitz symbol correspondence sequences are of Poisson (resp. anti-Poisson) type, but not of Bohr (resp. anti-Bohr) type (cf. Propositions 8.2.6 and 8.2.18). Similarly, both sequences in Example 8.2.19 are of Poisson type, but not of Bohr type.

The standard (resp. alternate) Stratonovich-Weyl symbol correspondence sequence is of pure-(resp. pure-anti)-Bohr type (cf. Propositions 8.2.5 and 8.2.18).

Example 8.2.27. If $f(l) = 1$ (resp. $f(l) = (-1)^l$), $\forall l \in \mathbb{N}$, any symbol correspondence sequence with $c_l^n = C(n, l) = f(l)g(n) \neq 0$, $\forall (n, l) \in \Delta^+(\mathbb{N}^2)$, is of Bohr (resp. anti-Bohr) type if $\lim_{n \to \infty} g(n) = 1$, and it is of pure-(resp. pure-anti)-Bohr type if $g(n) = 1 + o(n^{-1})$, $n \to \infty$.

Remark 8.2.28. The distinction between symbol correspondence sequences of Poisson type or Bohr type is manifest in the high-l-asymptotic dynamics of symbols, that is, asymptotic analysis of the dynamics of symbols when $l \to \infty$ as $j \to \infty$; in other words, highly oscillatory symbols and their twisted products.

Due to space and time constraints, in this monograph we shall not study the high-l-asymptotics of symbol correspondence sequences and twisted products, deferring this much harder study to a later opportunity.

8.3   Some Final Discussions and Considerations

8.3.1   Discussions on Our Approach

The attentive reader may, at this point, be asking the following question: since every twisted j-algebra of polynomial functions is isomorphic to the quantum algebra of finite $(n + 1)$-square matrices, while the classical Poisson algebra of smooth functions is infinite dimensional, how can one pose statements about the $n \to \infty$

limit of twisted products without addressing the problem of passing from finite to infinite matrices? In other words, we have so far been clever to avoid addressing this problem, but is this really justified? If so, in what sense?

In this respect, first we point out that various general works from different authors have studied the passage from finite matrices to infinite matrices and differential operators on infinite dimensional Hilbert spaces, which, as one should expect, is far from being a trivial problem, although it is facilitated in the case where all functions have compact support, as is the case of functions on the sphere.

But, by working directly with the symbols of the operators, we were able to bypass this subtle problem altogether, in the sense of looking at the $2j = n \to \infty$ limit as an asymptotic limit of twisted j-algebras. Let us expand on this point.

Knowing that $Poly_\mathbb{C}(S^2)_{\leq n} = \{Y_l^m\}_{-l \leq m \leq l \leq n}$ densely approximates $C_\mathbb{C}^\infty(S^2)$ as $n \to \infty$, we treat $1/n$, for $n \in \mathbb{N}$, as an asymptotic expansion parameter and look at the asymptotic expansions in this parameter of the expressions obtained for each sequence of twisted j-algebras, associated to each symbol correspondence sequence. As long as $1/n \neq 0$, we have $n \in \mathbb{N}$ and each $\vec{c}$-twisted j-algebra $(Poly_\mathbb{C}(S^2)_{\leq n}, \star_{\vec{c}}^n)$ is a finite-dimensional algebra isomorphic to the matrix algebra of the spin-j system. The case $1/n = 0$ is never really considered, just as ∞ is neither a real nor a natural number. Thus, for each $n \in \mathbb{N}$ and general $Y_{l_1}^{m_1}, Y_{l_2}^{m_2} \in Poly_\mathbb{C}(S^2)_{\leq n}$,

$$(Y_{l_1}^{m_1} \star_{\vec{c}}^n Y_{l_2}^{m_2} + Y_{l_2}^{m_2} \star_{\vec{c}}^n Y_{l_1}^{m_1})/2 - Y_{l_1}^{m_1} Y_{l_2}^{m_2} \neq 0 \,,$$

$$n(Y_{l_1}^{m_1} \star_{\vec{c}}^n Y_{l_2}^{m_2} - Y_{l_2}^{m_2} \star_{\vec{c}}^n Y_{l_1}^{m_1})/2 - i\{Y_{l_1}^{m_1}, Y_{l_2}^{m_2}\} \neq 0 \,.$$

However, the errors, i.e. differences from zero in the l.h.s. of these expressions, become smaller and smaller as n increases, *if and only if* we have $\forall l \leq n$ that $|c_l^n - 1|$ is either zero or becomes smaller and smaller as n increases keeping the l's fixed (similarly for considering $|c_l^n - (-1)^l|$ in the anti-Poisson case).

In other words, taking the sup-norm in the space of smooth functions on the sphere, $||f|| = sup(|f(\mathbf{n})|, \mathbf{n} \in S^2)$, then we can rewrite the above expressions as

$$||(Y_{l_1}^{m_1} \star_{\vec{c}}^n Y_{l_2}^{m_2} + Y_{l_2}^{m_2} \star_{\vec{c}}^n Y_{l_1}^{m_1})/2 - Y_{l_1}^{m_1} Y_{l_2}^{m_2}|| = S[\vec{c}]_{l_1,l_2}^{m_1,m_2}(n) \,, \quad (8.56)$$

$$||n(Y_{l_1}^{m_1} \star_{\vec{c}}^n Y_{l_2}^{m_2} - Y_{l_2}^{m_2} \star_{\vec{c}}^n Y_{l_1}^{m_1})/2 - i\{Y_{l_1}^{m_1}, Y_{l_2}^{m_2}\}|| = A[\vec{c}]_{l_1,l_2}^{m_1,m_2}(n) \,, \quad (8.57)$$

where $S[\vec{c}]_{l_1,l_2}^{m_1,m_2}$, $A[\vec{c}]_{l_1,l_2}^{m_1,m_2} : \mathbb{N} \to \mathbb{R}^+$ are sequences of nonnegative real numbers satisfying

$$\lim_{n \to \infty} S[\vec{c}]_{l_1,l_2}^{m_1,m_2}(n) = \lim_{n \to \infty} A[\vec{c}]_{l_1,l_2}^{m_1,m_2}(n) = 0 \,, \quad (8.58)$$

for general l_1, l_2, *if and only if* the sequence of characteristic numbers c_l^n satisfies the conditions of Theorem 8.2.21, cf. Eq. (8.53) (or Eq. (8.54) for anti-Poisson).

Therefore, if this asymptotic condition on the characteristic numbers c_l^n is satisfied, then we can say that the corresponding sequence of $\vec{c}$-twisted j-algebras approximates better and better the infinite-dimensional Poisson algebra of smooth functions on the sphere, knowing that for each $n \in \mathbb{N}$ we are only considering a finite-dimensional algebra, the $\vec{c}$-twisted j-algebra $(Poly_{\mathbb{C}}(S^2)_{\leq n}, \star_{\vec{c}}^n)$ which is isomorphic to the operator algebra of the corresponding spin-j quantum system.

In this respect, our approach is somewhat similar in spirit to the approach developed by Rieffel [57, 58] that uses the standard Berezin (and Toeplitz) symbol correspondences to show that the operator algebra of spin-j systems "converges in quantum Gromov-Hausdorff distance", as $n = 2j \to \infty$, to the algebra of continuous functions on the sphere (under pointwise product, i.e. concerning expressions like (8.56) only). Much of Rieffel's work centers on precisely defining and proving this metric convergence, for which he relies on the Berezin transform (see Definition 7.2.32). We note, however, that the standard Berezin (and Toeplitz) symbol correspondence sequences satisfy Theorem 8.2.21, meaning that the above necessary and sufficient condition on the characteristic numbers c_l^n is satisfied (and as a consequence, we have Remark 8.2.20 (ii), on the Berezin transform).

We should also point out, however, that the asymptotic relation obtained via Eqs. (8.56)–(8.58) under the condition of Theorem 8.2.21, between the sequence of operator algebras of spin-j systems and the Poisson algebra on S^2, is considerably simpler than any similar asymptotic relation involving the algebra of bounded operators on $L_{\mathbb{C}}^2(\mathbb{R})$ and the Poisson algebra on $\mathbb{R}^2$, for instance, where both algebras are infinite dimensional. In such cases, any sequence of finite dimensional algebras that aims at approximating the Poisson algebra asymptotically also needs to approximate the operator algebra at each stage, and controlling what happens to the latter (infinite-dimensional) kernel is another issue.

On the other hand, by working with sequences of spin-j operator algebras and their corresponding sequences of twisted j-algebras $(Poly_{\mathbb{C}}(S^2)_{\leq n}, \star_{\vec{c}}^n)$, we were able to discover an interesting phenomenon: although all $\vec{c}$-twisted j-algebras are isomorphic for each fixed $n = 2j \in \mathbb{N}$, any sequence of $\vec{c}$-twisted j-algebras that does not satisfy the condition that all $|c_l^n - 1|$ are zero or become smaller and smaller as n increases (or similarly for $|c_l^n - (-1)^l|$), will not provide a better and better approximation to the Poisson algebra on S^2. Again, this is an asymptotic statement that is verified for n larger and larger, though always finite.

But, as the generic condition on sequences of characteristic numbers c_l^n is the *failure* of the condition expressed by Theorem 8.2.21, cf. Eqs. (8.53) or (8.54), the interested reader could then wonder whether such generic failure of sequences of $\vec{c}$-twisted j-algebras to approximate the classical Poisson algebra can have measurable consequences. We now turn to this question.

8.3.2 "Empirical" Considerations

In standard quantum mechanics, the operators themselves are not measurable quantities, rather, measurable quantities are some expectation values which, for spin systems, are expressed by the Hilbert-Schmidt inner product (3.27), i.e.,

$$\langle P, Q \rangle = trace(P^* Q) . \tag{8.59}$$

In view of Eq. (6.13), the value at any point $\mathbf{n} \in S^2$ of the symbol $W_{\frac{j}{c}}(P)$ of any operator $P \in M_{\mathbb{C}}(n + 1)$ can be written in the form of Eq. (8.59) as an expectation value, for any symbol correspondence $W_{\frac{j}{c}}$.

Though most expectation values are quantum measurable quantities, the real ones are more easily so. Therefore, the value of any real symbol (i.e. symbol of any Hermitian operator) at any point on S^2 can be assumed to be a quantum measurable quantity for a spin system. On the other hand, the value of any real function at any point on S^2 is a classical measurable quantity for a spin system. Thus, under this assumption, the results of the previous section on the asymptotic dynamics of symbols acquire a definite measurable significance.

This is more clearly seen in the case of mapping-positive symbol correspondences. Remind that an operator $P = P^*$, with $trace(P) = 1$, is a pure-state if P is the projector onto a one-dimensional subspace and is a mixed-state if P is any convex combination of pure states. These operators are also called density operators or (generalized) states of the quantum system. Thus, when P is a state and $Q = Q^*$, Eq. (8.59) has a standard empirical meaning in quantum mechanics: it measures the expectation value of the observable Q in the state P.

Now, for mapping-positive symbol correspondences, the operator kernel K in (6.13) is itself a state, which is a pure state in the case of Berezin correspondences. And as shown explicitly in the case of the upper-middle-state and lower-middle-state correspondences (cf. (8.27)), general correspondence sequences obtained via a coherent (family of) state(s) are not of Poisson type. Therefore these symbols, whose values are standard quantum-measurable quantities, do not generally behave classically or near classically in the asymptotic limit of high spin (j) numbers.

However, for still more general symbol correspondences, even if the assumption on the quantum measurability of every real symbol is considerably relaxed, the results of the previous section still have far reaching measurability. To see this, let us focus on a simple instance, namely assume $H = H^*$, $A = A^*$, and set $Q = [H, A] = i\hbar \dot{A}$ (cf. (3.87)). Again let $P = P^*$, with $trace(P) = 1$, be a generalized state, but now the operator kernel K in (6.13) can be more general and is not necessarily a state. Then, with H being the Hamiltonian operator, equation

$$\langle P, \dot{A} \rangle = trace(P \dot{A}) = \frac{1}{i\hbar} trace(P[H, A]) \tag{8.60}$$

measures the time derivative of operator A in state P, or the time derivative of the expectation value of A in P (in the Heisenberg picture, see for instance [23]).

For any symbol correspondence $W_{\frac{j}{c}}$, the expectation value (8.60) can be written in either of the equivalent forms given by Eqs. (6.25) and (7.3),

$$\frac{n+1}{4\pi} \int_{S^2} \widetilde{W_{\stackrel{j}{c}}}(P)\, W_{\stackrel{j}{c}}(\dot{A})\, dS = \frac{n+1}{4\pi} \int_{S^2} W_{\stackrel{j}{c}}(P) \star_{\stackrel{n}{c}} W_{\stackrel{j}{c}}(\dot{A})\, dS$$

$$= \frac{n+1}{4\pi i \hbar} \int_{S^2} W_{\stackrel{j}{c}}(P) \star_{\stackrel{n}{c}} [W_{\stackrel{j}{c}}(H), W_{\stackrel{j}{c}}(A)]_{\star_{\stackrel{n}{c}}}\, dS \qquad (8.61)$$

$$= \frac{n+1}{4\pi i \hbar} \int_{S^2} \widetilde{W_{\stackrel{j}{c}}}(P)[W_{\stackrel{j}{c}}(H), W_{\stackrel{j}{c}}(A)]_{\star_{\stackrel{n}{c}}}\, dS \qquad (8.62)$$

where $\widetilde{W_{\stackrel{j}{c}}}$ is the symbol correspondence dual to $W_{\stackrel{j}{c}}$ (see Remark 6.2.25) and $[W_{\stackrel{j}{c}}(H), W_{\stackrel{j}{c}}(A)]_{\star_{\stackrel{n}{c}}}$ is the twisted commutator of $W_{\stackrel{j}{c}}(H)$ and $W_{\stackrel{j}{c}}(A)$.

Now, suppose that all above symbols have well defined $j \to \infty$ asymptotic limits of non highly-oscillatory type (decomposable into finite sums of spherical harmonics). In fact, let us also suppose that the symbol correspondence sequence of $W_{\stackrel{j}{c}}$ is of quasi-classical type (cf. Definition 8.2.17) and denote by p, h, a the asymptotic limits of $W_{\stackrel{j}{c}}(P)$, $W_{\stackrel{j}{c}}(H)$, $W_{\stackrel{j}{c}}(A)$, respectively. From the asymptotic self-dual property of symbol correspondence sequences of quasi-classical type (cf. Remark 8.2.20), the last expression (8.62) has asymptotic limit

$$\frac{1}{4\pi} \int_{S^2} p[h, a]_\infty dS, \qquad (8.63)$$

where

$$[h, a]_\infty = \lim_{n \to \infty} \frac{n+1}{i\hbar} [W_{\stackrel{j}{c}}(H), W_{\stackrel{j}{c}}(A)]_{\star_{\stackrel{n}{c}}}$$

(to directly relate this equation with the ones of the previous chapter, set $\hbar = 2$, reminding that, for spin systems, $\hbar$ is always treated as a real constant that can be omitted by proper scaling and the asymptotic limit is taken by letting $n \to \infty$).

On the other hand, from Hamilton-Poisson dynamics, the classical limit of (8.60) should be

$$\frac{1}{4\pi} \int_{S^2} p\{h, a\} dS, \qquad (8.64)$$

where $\{h, a\}$ is the Poisson bracket of h, a. However, if the quasi-classical symbol correspondence sequence of $W_{\stackrel{j}{c}}$ is not of Poisson type, as in Example 8.2.23 for instance, then $[h, a]_\infty \neq \{h, a\}$ and, in this case, the integrals (8.63) and (8.64) will in general not coincide.

Similar asymptotics of (8.60) can be performed for other kinds of limiting-type symbol correspondence sequences which are not of Poisson type, with similar conclusions. For instance, it is not too difficult to see that, for a characteristic-positive symbol correspondence sequence $W_{\stackrel{j}{c}}$ of pseudo-classical type (cf. Definition 8.2.17), if p, h, a are the asymptotic limits as before, then we have in general

$$\lim_{n\to\infty} \frac{n+1}{4\pi i \hbar} \int_{S^2} W_{\tilde{c}}^{j}(P) \star_{\tilde{c}}^{n} [W_{\tilde{c}}^{j}(H), W_{\tilde{c}}^{j}(A)]_{\star_{\tilde{c}}^{n}} dS \neq \frac{1}{4\pi} \int_{S^2} p\{h, a\} dS.$$

And this is typical for other instances of quantum expectation values, as well.

On the other hand, in view of Remark 8.2.24, every symbol correspondence sequence of pseudo-classical type (with characteristic numbers c_l^n) can in principle be mapped to a "renormalized" symbol correspondence sequence of Poisson type (with characteristic numbers χ_l^n), so that none of the measurable idiosyncrasies discussed above applies to the asymptotics of the "renormalized" symbols.

However, such a "renormalization" requires knowing $c_l^\infty = \lim_{n\to\infty} c_l^n$, $\forall l \in \mathbb{N}$, and this may not be the case, or it may be impractical to perform such an "asymptotic renormalization" on an already pre-established symbol correspondence sequence: just think again of a generic mapping-positive correspondence, in which case the symbols are expectation values in a specific coherent (family of) state(s).

Now, recalling that the standard Stratonovich-Weyl and Berezin symbol correspondence sequences are of Poisson type, it would be interesting to see whether the non-pure-Poisson property of standard Berezin symbols can have nontrivial measurability in low-l high-j-asymptotics. Furthermore, the identity (8.52) suggests that their high-l-asymptotics are quite different, in the standard Stratonovich-Weyl and standard Berezin (and Toeplitz) cases.

But this is a considerably harder question, which could perhaps be better addressed by using the integral formulations studied in Sect. 7.2. However, these would be much more useful, in this respect, if we had been able to obtain closed formulae for the integral trikernels under consideration. Therefore, further investigations in this direction could turn out to be profitable.

An alternative standpoint could be established by obtaining adequate asymptotic approximations for these trikernels which could be used in high-l asymptotic investigations. A possible approach to this goal is to introduce appropriate j-dependent scalings on the spheres so that, in the $n \to \infty$ limit, spherical patches tend to the symplectic plane and the spherical trikernels tend to well-known affine ones in small neighborhoods. In terms of the symmetry groups, this path leads to a contraction of the Lie algebra of $SU(2)$ to the Lie algebra of the Heisenberg group. Some work along these lines has been carried out in the context of the group $SU(1, 1)$, instead of $SU(2)$, where, by first performing a contraction of the Lie algebra and later performing a "quantized decontraction", some closed formulae for trikernels on the hyperbolic plane have been obtained [15]. One wonders, however, if the completely different topologies of $SU(2)$ and the Heisenberg group can allow for any useful outcome of this kind of procedure, in the spherical case.

Another approach is to use other asymptotic formulae for the Wigner $3jm$ and $6j$ symbols, that can be used to obtain high-l asymptotic approximations of the products and/or the trikernels. Particular formulae for these Wigner symbols are known in the asymptotic limit when the l's tend to ∞ linearly with j, that is, keeping all fractions l/j fixed. The respective nonuniform formulae have long been known, cf. [55], but uniform formulae are also known, cf. [3, 4]. Still another approach is to work with the integral formulae for the trikernels, which were obtained in Sect. 7.2.3, either to obtain asymptotic formulae for the trikernels themselves, or directly to the products of highly oscillatory functions.

Finally, we end this subsection with an important clarification of its context: at this point, all of the above "empirical" considerations are purely theoretical and whether any of these can eventually be actually observed in a real physical laboratory in some possible future, is at present totally unknown to us.

Chapter 9
Conclusion

Since the work of Bayen, Flato, Frondsal, Lichnerowicz and Sternheimer on deformation quantization [9], much emphasis has been placed on a class of problems initially known as quantization of Poisson manifolds. At first, the deformation quantization program, which started in [9] but was inspired by the much older work of Moyal [50], seemed to promise a definitive approach towards a precise mathematical relationship between quantum and classical mechanics in a unique and general setting. And soon, approaches to invariant deformation quantization were set forth, as the early work of Bayen and Frondsal [8] on the *formal* deformation quantization of the 2-sphere. Moreover, this promise of a general formalism showed itself stronger after the works of Fedosov [29] and Kontsevich [42], thus inspiring many to enlarge the program to ever more general settings.

However, as could have been clear from the start of the program, already for the case of affine symplectic spaces, the deformation quantization approach is not so well suited to handle highly oscillatory functions. These are common in some WKB semiclassical approximation of certain types of operators in ordinary quantum mechanics, particularly projectors or evolution operators in the Weyl representation (see the discussion in [60] and a related question in [59]).

Furthermore, the promise of a very general framework for quantization took a hard blow with the work of Rieffel [56], based on the work of Wassermann [80], which showed that, in the simple case of the homogeneous 2-sphere, any $SO(3)$-invariant "strict deformation quantization" of S^2 has to be isomorphic to some $SU(2)$-invariant finite matrix algebra, or some sequence of $SU(2)$-invariant matrix algebras in reverse order, i.e. of finite dimensions decreasing from infinity. Here, by *strict* deformation quantization, one should understand a closed associative noncommutative algebra in some function subspace of $C_{\mathbb{C}}(S^2)$, with all the required properties of a deformation quantization. Thus, in particular, equations analogous to (8.17)–(8.20) have to be satisfied (see [56] for more details; see also [18, 44]).

© Springer International Publishing Switzerland 2014
P. de M. Rios, E. Straume, *Symbol Correspondences for Spin Systems*,
DOI 10.1007/978-3-319-08198-4_9

In this way, once again one could not simply dismiss an understanding that the path from classical to quantum mechanics can be quite more subtle than straightforward, and quite more peculiar than generic.

On the other hand, the path from quantum to classical mechanics has often been thought to be unique, at least in principle. Despite the various methods of semiclassical approximation in affine mechanical systems, these have often been thought of as different approximations pertaining to an underlying unique limiting procedure. Thus, it is commonly believed that semiclassical approximations to the Weyl-Wigner formalism should, for instance, not be essentially different from semiclassical approximations to the coherent-state formalism, as the two approximations are commonly believed to be empirically equivalent.

The case of spin systems studied in this monograph shows that, on the contrary, the path from quantum to classical mechanics is very far from being unique. Different symbol correspondence sequences yield different semiclassical limits, when such a limit actually exists, which does not always happen.

Therefore, a generic symbol correspondence sequence defines a "quantization of S^2 in reverse order", i.e. from quantum to "classical", or better a sequence of "fuzzy spheres", in the sense of defining a sequence of function algebras satisfying Proposition 7.1.3, what we have called in this book a sequence of "$\vec{c}$-twisted j-algebras". However, generically this is not a reversed-order deformation of the classical sphere, in the sense that to be a reversed-order deformation of the classical sphere Eqs. (8.17)–(8.20) must also be satisfied.

Only a subclass of symbol correspondence sequences yields Poisson dynamics on S^2 in the asymptotic $n \to \infty$ limit. This subclass, the subclass of symbol correspondence sequences of Poisson (or anti-Poisson) type, realizes strict deformation quantizations of the classical two-sphere in reverse order. To this subclass belong the standard and the alternate Stratonovich-Weyl, as well as the standard and the alternate Berezin symbol correspondences, which are the spherical analogues of the Weyl-Wigner and the (standard) coherent-state representations of affine quantum mechanics, whose classical limits yield Poisson dynamics in affine symplectic space, at least for non-highly-oscillatory functions (cf. [60]).

Thus, it is important to emphasize that symbol correspondence sequences outside this subclass define symbolic dynamics on S^2 which in general will not be empirically equivalent to Poisson dynamics in the asymptotic $n \to \infty$ limit. This includes general cases of isometric symbol correspondence sequences and (mapping-positive) symbol correspondence sequences defined via coherent states, even those which are defined in simple and n-invariant ways, as is the case for the upper and lower-middle-state symbol correspondence sequences, for instance.

On the other hand, further investigations are in order, to assert the possibility of empirical distinctions within the subclass of symbol correspondence sequences of Poisson type. In this respect, we could benefit from a more detailed asymptotical comparison between the standard Stratonovich-Weyl and the standard Berezin (and Toeplitz) symbol correspondence sequences, particularly from the point of view of possible quantum measurability of their distinctions.

This could take the form of: (i) understanding possible measurable consequences of the higher order terms in the expansion (6.38) for the standard Berezin characteristic numbers, in low-l high-j asymptotics, or:

(ii) some substantial understanding of the high-l asymptotics of these correspondences, since (8.52) is an indication that the high-l asymptotical dynamics of the standard Berezin (and Toeplitz) symbol correspondence could be distinguishable from the high-l asymptotical dynamics of the standard Stratonovich-Weyl correspondence (or some other symbol correspondence sequence of Bohr type, arguably the "best asymptotical type", of which the standard and alternate Stratonovich-Weyl correspondences are the supreme prototypes).

Considerable help for (ii) could come from obtaining closed formulas for these trikernels, which we haven't yet been able to acquire, but a less ambitious goal would be deriving adequate asymptotical expressions for these trikernels which could be used for investigating (ii).

Finally, we close our present study with a more philosophical conclusion, perhaps the most important conclusion of this monograph.

It has long been recognized by many, mainly physicists but also mathematicians, that quantum mechanics "carries more information" or "is actually bigger" than classical mechanics, meaning that one cannot produce full quantum dynamics unambiguously, solely on the basis of classical data (the problem of strictly quantizing the sphere mentioned above being an instance of this general principle).

However, the abounding existence in spin systems of symbol correspondence sequences of non-Poisson type means that, in order to guarantee classical Poisson dynamics of spherical symbols, some limiting constraints must be placed upon the symbol correspondence sequences. In other words, classical information must be added to the quantum data, as well. This fact brings forth the realization, for spin systems, that a full consistent theory relating quantum and classical mechanics is actually bigger than either of these two theories alone.

Chapter 10
Appendix: Further Proofs

10.1 A Proof of Proposition 3.2.6

We shall derive the coupling rule (3.45) and its inversion (3.46) by formal reasoning with the Clebsch-Gordan coefficients. Starting from the formula (3.40), let $g \in SU(2)$ act on both sides, which yields (with $m = m_1 + m_2$)

$$\sum_{\mu_1,\mu_2} D^{j_1}_{\mu_1,m_1} D^{j_2}_{\mu_2,m_2} |j_1\mu_1 j_2\mu_2\rangle = \sum_j C^{j_1,j_2,j}_{m_1,m_2,m} \sum_\mu D^j_{\mu,m} |(j_1 j_2) j\mu\rangle . \qquad (10.1)$$

Now, substitute the expansion of type (3.44) for the coupled basis vector $|(j_1 j_2) j\mu\rangle$ in (10.1) and obtain (with $\mu = \mu'_1 + \mu'_2$)

$$\sum_{\mu_1,\mu_2} D^{j_1}_{\mu_1,m_1} D^{j_2}_{\mu_2,m_2} |j_1\mu_1 j_2\mu_2\rangle = \sum_j \sum_{\mu'_1,\mu'_2} C^{j_1,j_2,j}_{m_1,m_2,m} C^{j_1,j_2,j}_{\mu'_1,\mu'_2,\mu} D^j_{\mu,m} |j_1\mu'_1 j_2\mu'_2\rangle$$

$$(10.2)$$

$$= \sum_{\mu_1,\mu_2} \sum_j C^{j_1,j_2,j}_{m_1,m_2,m} C^{j_1,j_2,j}_{\mu_1,\mu_2,\mu} D^j_{\mu,m} |j_1\mu_1 j_2\mu_2\rangle ,$$

where the last expression follows from the second by the change of notation $\mu'_i \rightarrow \mu_i$. Since both sides of (10.2) are linear combinations of the uncoupled basis, corresponding coefficients are identical, consequently formula (3.45) must hold.

Next, by applying $g \in SU(2)$ to both sides of the formula (3.44),

$$\sum_\mu D^j_{\mu,m} |(j_1 j_2) j\mu\rangle = \sum_{m_1} C^{j_1,j_2,j}_{m_1,m_2,m} \sum_{\mu_1,\mu_2} D^{j_1}_{\mu_1,m_1} D^{j_2}_{\mu_2,m_2} |j_1\mu_1 j_2\mu_2\rangle$$

$$= \sum_{m_1} C^{j_1,j_2,j}_{m_1,m_2,m} \sum_{\mu_1,\mu_2} D^{j_1}_{\mu_1,m_1} D^{j_2}_{\mu_2,m_2} \sum_k C^{j_1,j_2,k}_{\mu_1,\mu_2,\mu} |(j_1 j_2) k\mu\rangle .$$

© Springer International Publishing Switzerland 2014

P. de M. Rios, E. Straume, *Symbol Correspondences for Spin Systems*,

DOI 10.1007/978-3-319-08198-4_10

Now, choosing $k = j$ and fixing the value of μ, comparison of the coefficient of $|(j_1 j_2) j\mu\rangle$ on both sides of the previous identity yields

$$D^j_{\mu,m} = \sum_{\mu_1,\mu_2} D^{j_1}_{\mu_1,m_1} D^{j_2}_{\mu_2,m_2} \sum_{m_1} C^{j_1,j_2,j}_{\mu_1,\mu_2,\mu} C^{j_1,j_2,j}_{m_1,m_2,m}$$

which is, in fact, the identity (3.46) since only terms with $\mu_2 = \mu - \mu_1$ can give a non-zero contribution.

10.2   A Proof of Proposition 3.3.12

The parity property for the product of operators, Proposition 3.3.12, follows straight-fowardly from the product rule for the coupled basis of operators, Corollary 3.3.22, and the symmetry properties of the Wigner $3jm$ and $6j$ symbols, as stated at the end of Chap. 2. However, it is possible to prove the parity property in an independent way, which highlights the large amounts of combinatorics that are encoded in the Wigner $3jm$ and $6j$ symbols. Thus, we now present this direct proof of Proposition 3.3.12, namely the parity property for the matrices

$$E(l,m) = (-1)^l \mu^n_{l,m} \mathbf{e}(l,m)$$

introduced in Sect. 3.3.2. This independent proof was worked out in collaboration with Nazira Harb.

First, we need some preliminary results. For greater clarity, we shall use the following notation:

$$A = J_+ \in \Delta(1), \ B = A^T = J_- \in \Delta(-1). \tag{10.3}$$

so that all matrices in $M_\mathbb{R}(n + 1)$ are expressible as linear combinations of monomials or "words" in the letters A and B. Observe that each monomial

$$P = A^{a_1} B^{b_1} A^{a_2} B^{b_2} \ldots A^{a_p} B^{b_p}, \ a_i \geq 0, b_i \geq 0 \tag{10.4}$$

is an m-subdiagonal matrix for some m in the range $-n \leq m \leq n$,

$$P = (x_1, x_2, \ldots, x_k)_m, \ k = n + 1 - |m| \tag{10.5}$$

with nonnegative entries x_i. We shall refer to m as the *weight* $\mu(P)$ of P, and consequently the monomial (10.4) has weight

$$m = \mu(P) = \sum a_i - \sum b_i.$$

In particular, $\mu(P^T) = -\mu(P)$, $\mu(A) = 1$, $\mu(B) = -1$, and diagonal matrices has weight zero. Moreover,

$$\mu(PQ) = \mu(P) + \mu(Q), \quad trace(P) \neq 0 \implies \mu(P) = 0.$$

We shall also compare a monomial with its *reverse* monomial, namely the reverse of X in (10.4) is by definition

$$P^{rev} = B^{b_p} A^{a_p} \ldots B^{b_2} A^{a_2} B^{b_1} A^{a_1} \tag{10.6}$$

Lemma 10.2.1. *A monomial matrix (10.4) and its reverse (cf. (10.6)) are related as follows;*

$$P = (x_1, x_2, \ldots, x_k)_m, \quad P^{rev} = (x_k, x_{k-1}, \ldots, x_1)_m.$$

In particular, $trace(P) = trace(P^{rev})$.

Proof. Define the *height* of the monomial in (10.4) to be the number $h(P) = \sum a_i + \sum b_i$. We shall prove the lemma by induction on the height (rather than weight). The lemma holds for monomials of height 1, namely A and B, which are their own reverse. For example,

$$A = (\alpha_1, \alpha_2, \ldots, \alpha_n)_1, \quad \alpha_1 = \alpha_n, \alpha_2 = \alpha_{n-1}, \ldots. \tag{10.7}$$

Now, assume the lemma holds for all monomials of height h, and let Y be a monomial of height $h + 1$. Then $Q = AP$ or BP, say $Q = AP$ where P is the m-subdiagonal matrix (10.5) and $0 \leq m < n$. By assumption, $P^{rev} = (x_k, x_{k-1}, \ldots, x_1)_m$ and we calculate

$$Q = (\alpha_1 x_2, \alpha_2 x_3, \ldots, \alpha_{k-1} x_k)_{m+1}, \quad Q^{rev} = P^{rev} A = (\alpha_{m+1} x_k, \alpha_{m+2} x_{k-1}, \ldots, \alpha_n x_2)_{m+1}.$$

By the symmetry of A illustrated in (10.7), the lemma also holds for Q. The case $Q = BP$ is similar and hence it is omitted. □

Now, we turn to the proof of the parity property. The two cases (i) and (ii) of Theorem 3.3.12 are similar, so let us choose case (i) and give a detailed proof, which amounts to showing the following inner product:

$$\langle E(l, m), [E(l_1, m_1), E(l_2, m_2)] \rangle$$

$$= \sum_{k=0}^{l-m} \sum_{i=0}^{l_1-m_1} \sum_{j=0}^{l_2-m_2} (-1)^{i+j+k} \binom{l-m}{k} \binom{l_1-m_1}{i} \binom{l_2-m_2}{j}$$

$$\cdot \{trace(A^k B^l A^{l-m-k} B^{l_1-m_1-i} A^{l_1} B^{l_2-m_2-j+i} A^{l_2} B^j)$$
$$- trace(A^k B^l A^{l-m-k} B^{l_2-m_2-j} A^{l_2} B^{l_1-m_1-i+j} A^{l_1} B^i)\}$$

vanishes when we assume $l \equiv l_1 + l_2 \pmod 2$ and $m = m_1 + m_2$. The vanishing of the inner product is immediate when $m \neq m_1 + m_2$ since each $E(l, m)$ is an m-subdiagonal matrix. Thus, for the proof, let us consider separately the two cases: either $l - m$ is odd or $l - m$ is even.

First, assume $l - m$ is odd and divide the summation over k into two sums:

$$\Sigma = \sum_{k=0}^{l-m} [\ldots] = \Sigma_1 + \Sigma_2 = \sum_{k=0}^{(l-m-1)/2} [\ldots] + \sum_{k=(l-m+1)/2}^{l-m} [\ldots]$$

In the second sum we make the substitution $(k, i, j) \rightarrow (t, r, s)$ by setting $t = l - m - k, r = l_1 - m_1 - i, s = l_2 - m_2 - j$; in particular

$$i + j + k = (l - m - t) + (l_1 - m_1 - r) + (l_2 - m_2 - s) \equiv r + s + t \pmod 2$$

because of the assumption $l \equiv l_1 + l_2$. Consequently, the second sum becomes

$$\Sigma_2 = \sum_{t=0}^{\frac{l-m-1}{2}} \sum_{r=0}^{l_1-m_1} \sum_{s=0}^{l_2-m_2} (-1)^{r+s+t} \binom{l-m}{t} \binom{l_1-m_1}{r} \binom{l_2-m_2}{s}$$

$$\cdot \{trace(A^{l-m-t} B^l A^t B^r A^{l_1} B^{l_1-m_1+s-r} A^{l_2} B^{l_2-m_2-s})$$

$$- trace(A^{l-m-t} B^l A^t B^s A^{l_2} B^{l_2-m_2-s+r} A^{l_1} B^{l_1-m_1-r})\},$$

and by the change of notation $(t, r, s) \rightarrow (k, i, j)$ in the expression Σ_2, we can write

$$\Sigma = \sum_{k=0}^{\frac{l-m-1}{2}} \sum_{i=0}^{l_1-m_1} \sum_{j=0}^{l_2-m_2} (-1)^{i+j+k} \binom{l-m}{k} \binom{l_1-m_1}{i} \binom{l_2-m_2}{j} \cdot \{\ldots\}$$

where

$$\{\ldots\} = \{trace(A^k B^l A^{l-m-k} B^{l_1-m_1-i} A^{l_1} B^{l_2-m_2-j+i} A^{l_2} B^j)$$

$$- trace(A^k B^l A^{l-m-k} B^{l_2-m_2-j} A^{l_2} B^{l_1-m_1-i+j} A^{l_1} B^i)$$

$$+ trace(A^{l-m-k} B^l A^k B^i A^{l_1} B^{l_1-m_1+j-i} A^{l_2} B^{l_2-m_2-j})$$

$$- trace(A^{l-m-k} B^l A^k B^j A^{l_2} B^{l_2-m_2-j+i} A^{l_1} B^{l_1-m_1-i})\}$$

$$= \{trace(A^k B^l A^{l-m-k} B^{l_1-m_1-i} A^{l_1} B^{l_2-m_2-j+i} A^{l_2} B^j)$$

$$- trace(B^j A^{l_2} B^{l_2-m_2-j+i} A^{l_1} B^{l_1-m_1-i} A^{l-m-k} B^l A^k)$$

$$+ trace(B^i A^{l_1} B^{l_1-m_1+j-i} A^{l_2} B^{l_2-m_2-j} A^{l-m-k} B^l A^k)$$

$$- trace(A^k B^l A^{l-m-k} B^{l_2-m_2-j} A^{l_2} B^{l_1-m_1-i+j} A^{l_1} B^i).$$

To obtain the last expression of {...} we have rearranged the four trace terms of {...} in the new order $1, 4, 3, 2$, and we have also made use of the cyclic property of the trace. Now, it follows from Lemma 10.2.1 that the expression {...} vanishes identically, for each triple (k, i, j) of indices.

Next, if $l - m$ is an even integer, we break the sum Σ over k into two sums Σ_1 and Σ_2. In the first sum, $k = 0, 1 \dots, \frac{l-m}{2} - 1$, plus the first trace term of {...} in (10.4) for $k = \frac{l-m}{2}$ (and summation over i, j, of course). In the second sum, $k = \frac{l-m}{2} + 1, \dots, l - m$, plus the second trace term for $k = \frac{l-m}{2}$. Then the proof of the vanishing of Σ follows analogously, and this completes the proof of property (i). Property (ii) is proven analogously.

Finally, to complete the proof of Theorem 3.3.12 it remains to show that the product $E(l_1, m_1) E(l_2, m_2)$ is a linear combination of terms $E(l, m)$ with $|l_1 - l_2| \le l \le l_1 + l_2$.

But, in the linear expansion of the product, a typical term $E(l, m)$ belongs to the matrix subspace $M_{\mathbb{C}}(\varphi_l)$. The operator $A = J_+$ acts as the derivation ad_A on matrices and leaves the subspace invariant, so by repeated application the term $E(l, m)$ is mapped to non-zero multiples of $E(l, m')$ with $|m'| \le l$. In particular, if the expansion has a term $E(l, m)$ with $l > l_1 + l_2$, say l is maximal, application of the operator A will map the expansion to a non-zero multiple of $E(l, l)$. On the other hand, the above product is an m-subdiagonal matrix and the action of A yields m'-subdiagonal matrices with m' at most equal to $l_1 + l_2$. This is a contradiction.

Next, let us assume $l_1 \ge l_2$, and suppose the expansion has the term $E(l, m)$ where l lies in the range $0 \le l < l_1 - l_2$. Application of A to this term can only yield terms $E(l, m')$ with $m' \le l$. However, application of A to the product also yields the term $E(l_1, l_1) E(l_2, m_2)$, which is m'-subdiagonal with $m' = l_1 + m_2$. On the other hand,

$$m' = l_1 + m_2 \ge l_1 - l_2 > l$$

and this is a contradiction.

10.3 A Proof of Proposition 3.3.24

Proposition 3.3.24 follows straight from the explicit formulae (3.108)–(3.109). However, it is interesting to see how it can be obtained directly from the general Eq. (3.103) defining the Wigner $6j$ symbol, as shown below.

Thus, we start from the following formula, which is a particular case of (3.103):

$$\begin{Bmatrix} l_1 & l_2 & l_3 \\ j & j & j \end{Bmatrix} = \sum (-1)^{3j + \delta + \epsilon + \phi} \qquad (10.8)$$

$$\cdot \begin{pmatrix} l_1 & l_2 & l_3 \\ \alpha & \beta & \gamma \end{pmatrix} \begin{pmatrix} l_1 & j & j \\ \alpha & \epsilon & -\phi \end{pmatrix} \begin{pmatrix} j & l_2 & j \\ -\delta & \beta & \phi \end{pmatrix} \begin{pmatrix} j & j & l_3 \\ \delta & -\epsilon & \gamma \end{pmatrix}$$

where, again, the sum is taken over all possible values of $\alpha, \beta, \gamma, \delta, \epsilon, \phi$, and only three of these are independent. Therefore,

$$\begin{Bmatrix} l_1 & l_3 & l_2 \\ j & j & j \end{Bmatrix} = \sum (-1)^{3j+\delta'+\epsilon'+\phi'} \tag{10.9}$$

$$\cdot \begin{pmatrix} l_1 & l_3 & l_2 \\ \alpha' & \beta' & \gamma' \end{pmatrix} \begin{pmatrix} l_1 & j & j \\ \alpha' & \epsilon' & -\phi' \end{pmatrix} \begin{pmatrix} j & l_3 & j \\ -\delta' & \beta' & \phi' \end{pmatrix} \begin{pmatrix} j & j & l_2 \\ \delta' & -\epsilon' & \gamma' \end{pmatrix} .$$

Using (3.95) and re-naming $\alpha' = \alpha, \ \beta' = \gamma, \ \gamma' = \beta$, from (10.9) we get

$$\begin{Bmatrix} l_1 & l_3 & l_2 \\ j & j & j \end{Bmatrix} = \sum (-1)^{3j+\delta'+\epsilon'+\phi'+l_1+2l_2+2l_3+4j} \tag{10.10}$$

$$\cdot \begin{pmatrix} l_1 & l_2 & l_3 \\ \alpha & \beta & \gamma \end{pmatrix} \begin{pmatrix} l_1 & j & j \\ \alpha & \epsilon' & -\phi' \end{pmatrix} \begin{pmatrix} j & j & l_3 \\ -\delta' & \phi' & \gamma \end{pmatrix} \begin{pmatrix} j & l_2 & j \\ \delta' & \beta & -\epsilon' \end{pmatrix} .$$

Renaming $\delta' = -\delta, \ \epsilon' = -\phi, \ \phi' = -\epsilon$, from (10.10) we get

$$\begin{Bmatrix} l_1 & l_3 & l_2 \\ j & j & j \end{Bmatrix} = \sum (-1)^{3j-\delta-\epsilon-\phi+l_1+2l_2+2l_3+4j} \tag{10.11}$$

$$\cdot \begin{pmatrix} l_1 & l_2 & l_3 \\ \alpha & \beta & \gamma \end{pmatrix} \begin{pmatrix} l_1 & j & j \\ \alpha & -\phi & \epsilon \end{pmatrix} \begin{pmatrix} j & j & l_3 \\ \delta & -\epsilon & \gamma \end{pmatrix} \begin{pmatrix} j & l_2 & j \\ -\delta & \beta & \phi \end{pmatrix} .$$

Again using (3.95), from (10.11) we get

$$\begin{Bmatrix} l_1 & l_3 & l_2 \\ j & j & j \end{Bmatrix} = \sum (-1)^{3j-\delta-\epsilon-\phi+2l_1+2l_2+2l_3+6j} \tag{10.12}$$

$$\cdot \begin{pmatrix} l_1 & l_2 & l_3 \\ \alpha & \beta & \gamma \end{pmatrix} \begin{pmatrix} l_1 & j & j \\ \alpha & \epsilon & -\phi \end{pmatrix} \begin{pmatrix} j & l_2 & j \\ -\delta & \beta & \phi \end{pmatrix} \begin{pmatrix} j & j & l_3 \\ \delta & -\epsilon & \gamma \end{pmatrix} .$$

But $(-1)^{3j-\delta-\epsilon-\phi+2l_1+2l_2+2l_3+6j} = (-1)^{3j+\delta+\epsilon+\phi}(-1)^{2(l_1+l_2+l_3+2j)}(-1)^{2(j-\delta-\epsilon-\phi)}$ and $(-1)^{2(l_1+l_2+l_3+2j)} = (-1)^{2(j-\delta-\epsilon-\phi)} = 1$, so the values of (10.12) and (10.8) are identical.

Similarly, permutation of any other two columns in (10.8) leaves the value invariant.

10.4 A Proof of Proposition 4.2.8

The calculation for the decomposition of the Poisson bracket

$$\left\{ Y_{l_1}^{m_1}, Y_{l_2}^{m_2} \right\} = \frac{(-1)^{l_1+l_2}}{\lambda_{l_1,m_1}\lambda_{l_2,m_2}} \left\{ J_-^{l_1-m_1}(x+iy)^{l_1}, J_-^{l_2-m_2}(x+iy)^{l_2} \right\} \tag{10.13}$$

can be considerably simplified by the appropriate choices of coordinates on $\mathbb{R}^3$, perhaps also complex coordinates since the functions are complex. Thus, in addition to (x, y, z) and spherical polar coordinates

$$(\rho, \varphi, \theta) : x = \rho \sin \varphi \cos \theta, \, y = \rho \sin \varphi \sin \theta, z = \cos \varphi,$$

following [30] we shall also express the various vector fields (or infinitesimal operators) in terms of the coordinate system

$$(u, v, z) : u = x + iy, \, v = x - iy, \, z = z \tag{10.14}$$

(the main difficulty with using only spherical polar coordinates for this calculation lies in handling the derivatives of the associated Legendre polynomials).

Now, via the action of $SO(3)$ on $\mathbb{R}^3$ the angular momentum operators J_k act as derivations of functions, yielding the following (complex-valued) vector fields

$$J_1 = i\left(z\frac{\partial}{\partial y} - y\frac{\partial}{\partial z}\right) = i\left(\sin\theta\frac{\partial}{\partial\varphi} + \cot\varphi\cos\theta\frac{\partial}{\partial\theta}\right),$$

$$J_2 = i\left(x\frac{\partial}{\partial z} - z\frac{\partial}{\partial x}\right) = i\left(-\cos\theta\frac{\partial}{\partial\varphi} + \cot\varphi\sin\theta\frac{\partial}{\partial\theta}\right),$$

$$J_3 = i\left(y\frac{\partial}{\partial x} - x\frac{\partial}{\partial y}\right) = u\frac{\partial}{\partial u} - v\frac{\partial}{\partial v} = -i\frac{\partial}{\partial\theta}, \tag{10.15}$$

$$J_+ = J_1 + iJ_2 = 2z\frac{\partial}{\partial v} - u\frac{\partial}{\partial z} = e^{i\theta}\left(\frac{\partial}{\partial\varphi} + i\cot\varphi\frac{\partial}{\partial\theta}\right),$$

$$J_- = J_1 - iJ_2 = -2z\frac{\partial}{\partial u} + v\frac{\partial}{\partial z} = e^{-i\theta}\left(-\frac{\partial}{\partial\varphi} + i\cot\varphi\frac{\partial}{\partial\theta}\right),$$

which are also tangential to the unit sphere $S^2 = (\rho = 1)$. Let us also express the coordinate vector fields of the system (10.14) in terms of spherical coordinates

$$\frac{\partial}{\partial u} = \frac{1}{2}\left(\sin\varphi e^{-i\theta}\frac{\partial}{\partial\rho} + \cos\varphi e^{-i\theta}\frac{1}{\rho}\frac{\partial}{\partial\varphi} - \frac{ie^{-i\theta}}{\rho\sin\varphi}\frac{\partial}{\partial\theta}\right),$$

$$\frac{\partial}{\partial v} = \frac{1}{2}\left(\sin\varphi e^{i\theta}\frac{\partial}{\partial\rho} + \cos\varphi e^{i\theta}\frac{1}{\rho}\frac{\partial}{\partial\varphi} + \frac{ie^{i\theta}}{\rho\sin\varphi}\frac{\partial}{\partial\theta}\right), \tag{10.16}$$

$$\frac{\partial}{\partial z} = \cos\varphi\frac{\partial}{\partial\rho} - \sin\varphi\frac{1}{\rho}\frac{\partial}{\partial\varphi}.$$

In particular, along the sphere S^2 these operators also have a component in the normal direction $\frac{\partial}{\partial\rho}$, which is simply ignored when we calculate the Poisson bracket (4.8) on S^2. The following lemma turns out to be very useful.

Lemma 10.4.1. *The Poisson bracket* $\{F, G\}$ *on the 2-sphere can be expressed by the formula*

$$i\{F, G\} = (\frac{\partial F}{\partial u})(J_+G) + (\frac{\partial F}{\partial v})(J_-G) + (\frac{\partial F}{\partial z})(J_3G). \tag{10.17}$$

Proof. It is straightforward to calculate the right-hand side of the identity in terms of the coordinates (φ, θ), using the expressions (10.15) and (10.16). Then one arrives at the expression (4.8) multiplied by i. $\square$

The spherical harmonics Y_l^m are generated by the successive application of the operator J_- to the monomial u^l, and calculation of their Poisson bracket (10.13) amounts to applying operator products of type $\frac{\partial}{\partial\varphi}J_-^k$ and $\frac{\partial}{\partial\theta}J_-^k$ to u^l. However, since the commutation relations between J_-^k and $\frac{\partial}{\partial\varphi}$ or $\frac{\partial}{\partial\theta}$ are rather intricate, the coordinates (10.14) suggest themselves as more suitable for calculation of the bracket of these particular functions.

In fact, the operator $\frac{\partial}{\partial u}$ commutes with J_-. This is, indeed, the motivation for the above lemma, cf. also formula (B14) in [30]. Using the expressions (10.15) the following commutation identities are easily proved by induction;

$$\frac{\partial}{\partial v}J_-^k = J_-^k\frac{\partial}{\partial v} + kJ_-^{k-1}\frac{\partial}{\partial z} - k(k-1)J_-^{k-2}\frac{\partial}{\partial u},$$

$$\frac{\partial}{\partial z}J_-^k = J_-^k\frac{\partial}{\partial z} - 2kJ_-^{k-1}\frac{\partial}{\partial u}, \quad J_3J_-^k = J_-^kJ_3 - kJ_-^k.$$

Consequently,

$$\frac{\partial}{\partial u}Y_l^m = -\frac{1}{2}\sqrt{\frac{2l+1}{2l-1}}\sqrt{(l+m)(l+m-1)}Y_{l-1}^{m-1},$$

$$\frac{\partial}{\partial v}Y_l^m = \frac{1}{2}\sqrt{\frac{2l+1}{2l-1}}\sqrt{(l-m)(l-m-1)}Y_{l-1}^{m+1},$$

$$\frac{\partial}{\partial z}Y_l^m = \sqrt{\frac{2l+1}{2l-1}}\sqrt{(l+m)(l-m)}Y_{l-1}^m,$$

$$J_-Y_l^m = \sqrt{(l+m)(l-m+1)}Y_l^{m-1},$$

$$J_+Y_l^m = \sqrt{(l-m)(l+m+1)}Y_l^{m+1}, \quad J_3Y_l^m, = mY_l^m$$

and substitution into formula (10.17) yields

$$i\left\{Y_{l_1}^{m_1}, Y_{l_2}^{m_2}\right\} = \frac{1}{2}\sqrt{\frac{2l_1+1}{2l_1-1}}$$

$$\sqrt{(l_1-m_1)(l_1-m_1-1)(l_2+m_2)(l_2-m_2+1)}Y_{l_1-1}^{m_1+1}Y_{l_2}^{m_2-1}$$

$$-\frac{1}{2}\sqrt{\frac{2l_1+1}{2l_1-1}}$$

$$\sqrt{(l_1+m_1)(l_1+m_1-1)(l_2-m_2)(l_2+m_2+1)}Y_{l_1-1}^{m_1-1}Y_{l_2}^{m_2+1}$$

$$+\sqrt{\frac{2l_1+1}{2l_1-1}}\, m_2\sqrt{(l_1-m_1)(l_1+m_1)}Y_{l_1-1}^{m_1}Y_{l_2}^{m_2}$$

Combining this with the product formula (4.44) we arrive at

$$i\left\{Y_{l_1}^{m_1}, Y_{l_2}^{m_2}\right\} = \sum_{\substack{l=|l_1-l_2|+1 \\ l \equiv l_1+l_2-1}}^{l_1+l_2-1} \sqrt{\frac{(2l_1+1)(2l_2+1)}{2l+1}}\, C_{0,0,0}^{l_1-1,l_2,l}\, P_{m_1,m_2,m}^{l_1-1,l_2,l}\, Y_l^m \qquad (10.18)$$

where $m = m_1 + m_2$, and

$$P_{m_1,m_2,m}^{l_1-1,l_2,l} = \frac{1}{2}\sqrt{(l_1-m_1)(l_1-m_1-1)(l_2+m_2)(l_2-m_2+1)}C_{m_1+1,m_2-1,m}^{l_1-1,l_2,l}$$

$$-\frac{1}{2}\sqrt{(l_1+m_1)(l_1+m_1-1)(l_2-m_2)(l_2+m_2+1)}C_{m_1-1,m_2+1,m}^{l_1-1,l_2,l}$$

$$+ m_2\sqrt{(l_1-m_1)(l_1+m_1)}C_{m_1,m_2,m}^{l_1-1,l_2,l}. \qquad (10.19)$$

(We mention that Eqs. (10.18)–(10.19) can be put in a more symmetric form by writing similar equations for $\left\{Y_{l_2}^{m_2}, Y_{l_1}^{m_1}\right\}$ and using the skew symmetry of the Poisson bracket to write $\left\{Y_{l_1}^{m_1}, Y_{l_2}^{m_2}\right\} = \frac{1}{2}\left(\left\{Y_{l_1}^{m_1}, Y_{l_2}^{m_2}\right\} - \left\{Y_{l_2}^{m_2}, Y_{l_1}^{m_1}\right\}\right)$.)

Now, in order to obtain (4.51) we use equivariance under the group action. Let us introduce the symbol

$$K_{m_1,m_2,m}^{l_1,l_2,l} = \sqrt{\frac{(2l_1+1)(2l_2+1)}{2l+1}}\, C_{0,0,0}^{l_1-1,l_2,l}\, P_{m_1,m_2,m}^{l_1-1,l_2,l}. \qquad (10.20)$$

From Eq. (4.45), we have on the one hand

$$\{Y^g_{l_1,m_1}, Y^g_{l_2,m_2}\} = \sum_{\mu_1,\mu_2} D^{l_1}_{\mu_1,m_1}(g) D^{l_2}_{\mu_2,m_2}(g)\{Y_{l_1,\mu_1}, Y_{l_2,\mu_2}\}$$

$$= \sum_{\mu_1,\mu_2,l} D^{l_1}_{\mu_1,m_1}(g) D^{l_2}_{\mu_2,m_2}(g) K^{l_1,l_2,l}_{\mu_1,\mu_2,\mu} Y_{l,\mu} , \tag{10.21}$$

where we have used (10.18) and (10.20), with summation in l under the appropriate restriction indicated in (10.18). On the other hand,

$$\{Y^g_{l_1,m_1}, Y^g_{l_2,m_2}\} = \sum_{l} K^{l_1, l_2, l}_{m_1,m_2,m} Y^g_{l,m}$$

$$= \sum_{l,\mu} D^l_{\mu,m}(g) K^{l_1, l_2, l}_{m_1,m_2,m} Y_{l,\mu} . \tag{10.22}$$

But by the coupling rule, Eq. (3.45), we can rewrite Eq. (10.21) as

$$\{Y^g_{l_1,m_1}, Y^g_{l_2,m_2}\} = \sum_{\mu_1,\mu_2,l,l'} C^{l_1,l_2,l'}_{\mu_1,\mu_2,\mu'} C^{l_1,l_2,l'}_{m_1,m_2,m'} D^{l'}_{\mu',m'}(g) K^{l_1, l_2, l}_{\mu_1,\mu_2,\mu} Y_{l,\mu} . \tag{10.23}$$

Then, using the orthonormality relations (3.43) of the Clebsch-Gordan coefficients, we conclude that "solutions" of (10.22) = (10.23) are given by

$$K^{l_1, l_2, l}_{m_1,m_2,m} = F(l_1, l_2, l)\, C^{l_1,l_2,l}_{m_1,m_2,m} , \tag{10.24}$$

where, in principle, F could be any function of l_1, l_2, l. However, writing

$$F(l_1, l_2, l) = \sqrt{\frac{(2l_1 + 1)(2l_2 + 1)}{2l + 1}}\, P(l_1, l_2, l)$$

and substituting (10.24) into (10.20), we see that the function $P(l_1, l_2, l)$ is determined by

$$C^{l_1-1,l_2,l}_{0,0,0}\, P^{l_1-1,l_2,l}_{m_1,m_2,m} = C^{l_1, l_2, l}_{m_1,m_2,m}\, P(l_1, l_2, l) . \tag{10.25}$$

Now, first we note that the l.h.s. of (10.25) vanishes if $l_1 + l_2 + l$ is even (cf. (4.50)), and therefore

$$P(l_1, l_2, l) \equiv 0 , \text{ if } l_1 + l_2 + l \text{ is even}, \tag{10.26}$$

which agrees with the sum in (4.51) being restricted to $l \equiv l_1 + l_2 - 1 \pmod 2$.

Second, we note that Eq. (4.2.6) must hold for any values of m_1 and m_2; thus, in particular, for $m = m_1 + m_2 = l$ and $m_1 = l_1$ we have that

$$C^{l_1-1,l_2,l}_{0,0,0}\, P^{l_1-1,l_2,l}_{l_1,l-l_1,l} = C^{l_1, l_2, l}_{l_1,l-l_1,l}\, P(l_1, l_2, l) . \tag{10.27}$$

The Clebsch-Gordan coefficient $C_{0,0,0}^{l_1-1,l_2,l}$ has the closed formula given by Eq. (4.49), with l_1 replaced by $l_1 - 1$, but similarly, the Clebsch-Gordan coefficients $C_{l_1,l-l_1,l}^{l_1,\,l_2,\,l}$ also have a well-known simple closed formula (cf. [74]):

$$C_{l_1,l-l_1,l}^{l_1,l_2,l} = \sqrt{\frac{(2l_1)!(2l+1)!}{(l_1+l_2+l+1)!(l_1-l_2+l)!}}. \tag{10.28}$$

Then, Eqs. (10.27) and (10.19), together with Eqs. (4.49) and (10.28) straightforwardly yield Eq. (4.52), when $l_1 + l_2 + l$ is odd.

10.5   A Proof of Proposition 6.2.34

The formula (6.37) in Proposition 6.2.34 follows directly from Eq. (6.36) and the explicit formulae (3.47)–(3.49) for the Clebsch-Gordan coefficients. Nonetheless, it is interesting to see how it can be obtained more directly and independently of these formulae.

To begin with, for $l = 1$ we have by (3.15) and (3.68)

$$\mathbf{e}(1,0) = \frac{1}{\beta_{1,1}}[J_-, \mathbf{e}(1,1)] = \frac{1}{\beta_{1,1}\mu_1}[J_+, J_-] = \frac{\sqrt{2}}{\mu_1}J_3.$$

and consequently

$$b_1^n = \sqrt{\frac{n}{n+2}}.$$

We shall work out the general formula

$$\mathbf{e}^j(l,0)_{1,1} = \frac{1}{\mu_l\sqrt{(2l)!}}x_1^2 x_2^2 \dots x_l^2, \quad \text{where } x_k = \sqrt{k(n-k+1)}, \tag{10.29}$$

and then, formula (6.37) follows immediately from

$$b_l^n = \sqrt{\frac{n+1}{2l+1}}\frac{x_1^2 x_2^2 \dots x_l^2}{\sqrt{(2l)!}\mu_l}. \tag{10.30}$$

So, let us focus on this formula, which can be proved by induction on l. The underlying calculations are simpler by working with the matrices $E(l,0)$ rather than the normed matrices $\mathbf{e}(l,0)$, so let us illustrate the idea by taking $l = 3$ and formally calculate subdiagonal matrices

$$A = (x_1, x_2, \ldots, x_n)_1, \, B = (x_1, x_2, \ldots, x_n)_{-1},$$

$$A^2 = (y_1, y_2, \ldots, y_{n-1})_2, \, y_k = x_k x_{k+1},$$

$$A^3 = (z_1, z_2, \ldots, z_{n-2})_3, \, z_k = x_{k+2} y_k = x_k x_{k+1} x_{k+2},$$

$$E(3,2) = [B, A^3] = (u_1, u_2, \ldots, u_{n-1})_2, u_1 = -x_3 z_1 = -x_1 x_2 x_3^2,$$

$$E(3,1) = [B, [B, A^3]] = (v_1, v_2, \ldots, v_n)_1, \, v_1 = -x_2 u_1 = x_1 x_2^2 x_3^2,$$

$$E(3,0) = [B, [B, [B, A^3]]] = (w_1, w_2, \ldots, w_{n+1})_0, \, w_1 = -x_1 v_1 = -x_1^2 x_2^2 x_3^2.$$

Thus, the first entry of the diagonal matrix $E(l,0)$ is seen to be

$$E(l,0)_{1,1} = (-1)^l x_1^2 x_2^2 \ldots x_l^2,$$

and on the other hand (cf. Sect. 3.3.2)

$$\mathbf{e}(l,0) = \frac{(-1)^l}{\mu_{l,0}^n} E(l,0), \tag{10.31}$$

where by (3.77)

$$\mu_{l,0}^n = \mu_l \sqrt{l!} \sqrt{(2l)(2l-1)(2l-2) \ldots (l+1)} = \mu_l \sqrt{(2l)!}$$

is the norm of $E(l,0)$. Now, formula (10.30) follows from (6.35) and (10.31).

10.6   A Proof of Proposition 6.2.54

We have to show that $p_l^n(1/2) \neq 0$, $\forall n = 2j \in \mathbb{N}$, $\forall l = 1, 2, \cdots, n$. But

$$p_l^n(1/2) = (b_l^n([j + 1/2]) + b_l^n([j + 1]))/2, \tag{10.32}$$

where $b_l^n([j + 1/2])$ and $b_l^n([j + 1])$ are the characteristic numbers of the Berezin pre-symbol maps defined by projectors $\Pi_k = \Pi_{[j+1/2]}$ and $\Pi_k = \Pi_{[j+1]}$, respectively.

Starting with the case $[j + 1/2]$, when j is a half-integer, $k = [j + 1/2] = j + 1/2$. In this case, $m = j - k + 1 = 1/2$. When j is an integer, $k = [j + 1/2] = j$. In this case, $m = j - k + 1 = 1$. Therefore, from (6.47) we have that

$$b_l^n([j + 1/2]) = (-1)^{j-1/2} \sqrt{\frac{n+1}{2l+1}} C_{1/2,-1/2,0}^{j,\,j,\,l}, \, [j + 1/2] = j + 1/2,$$

$$= (-1)^{j-1} \sqrt{\frac{n+1}{2l+1}} C_{1,-1,0}^{j,\,j,\,l}, \, [j + 1/2] = j. \tag{10.33}$$

The case $[j + 1]$ only differs from the case $[j + 1/2]$ when j is an integer and then $k = [j + 1] = j + 1$, so that $m = j - k + 1 = 0$. Therefore,

$$b_l^n([j + 1]) = b_l^n([j + 1/2]) \ , \ [j + 1] = j + 1/2 \,, \tag{10.34}$$

$$= (-1)^j \sqrt{\frac{n + 1}{2l + 1}} C_{0,0,0}^{j, \, j, \, l} \ , \ [j + 1] = j + 1 \,.$$

From (10.33) and (10.34) we arrive at the following explicit expressions for the characteristic numbers in (10.32), namely

$$p_l^n(1/2) = \begin{cases} (-1)^{j-1/2} \sqrt{\frac{n+1}{2l+1}} C_{1/2,-1/2,0}^{j,j,l} = b_l^n(j + 1/2) \ , \ j \ \text{half-integral} \\[2mm] (-1)^j \sqrt{\frac{n+1}{2l+1}} \left(\frac{1}{2}(-C_{1,-1,0}^{j,j,l} + C_{0,0,0}^{j,j,l}) \right) \ , \ j \ \text{integral.} \end{cases}$$

In both cases for $p_l^n(1/2)$ as above (for n even or odd), there are different formulas for the Clebsch-Gordan coefficients in (10.33), depending on whether l is even or odd, and they are related by well-known recursive relations (cf. Section 8.6 in [72]) which enable us to express $p_l^n(1/2)$ as a certain multiple of a Clebsch-Gordan coefficient of type $C_{0,0,0}^{l_1,l_2,l_3}$ with $(l_1 + l_2 + l_3)$ even, so that we can use (4.49).

So let us consider all four cases separately. For the case n odd (j half-integer), we obtain from formulas (18) and (19) in section 8.6.3 of [74], that

$$C_{1/2,-1/2,0}^{j, \, j, \, l} = \frac{\sqrt{(n - l)(n + l + 1)}}{n + 1} C_{0, \ 0, \ 0}^{j-1/2, j-1/2, l} \ , \ n \ \text{odd,} \ l \ \text{even,} \tag{10.35}$$

$$C_{1/2,-1/2,0}^{j, \, j, \, l} = \frac{\sqrt{l(l + 1)}}{n + 1} C_{0, \ 0, \ 0}^{j+1/2, j-1/2, l} \ , \ n \ \text{odd,} \ l \ \text{odd,} \tag{10.36}$$

while the case n even (j integer) and l even is obtained from formula (7) in section 8.6.2 of [74] as

$$C_{1,-1,0}^{j, \, j, \, l} = \frac{2l(l + 1) - n(n + 2)}{n(n + 2)} C_{0,0,0}^{j,j,l} \ , \ n \ \text{even,} \ l \ \text{even,} \tag{10.37}$$

and in the case n even l odd, a closed formula for $C_{1,-1,0}^{j, \, j, \, l}$ is obtained by combining formulas (9), (18) and (19) of section 8.6.3 in [74], yielding

$$C_{1,-1,0}^{j, \, j, \, l} = \frac{2\sqrt{(n - l)(n + l + 1)l(l + 1)}}{n(n + 2)} C_{0, \ 0, \ 0}^{j,j-1,l} \ , \ n \ \text{even,} \ l \ \text{odd.} \tag{10.38}$$

We also recall that in the j integral (n even) case for (10.34), the only nontrivial case is when l is even. Now, combining (10.32)–(10.38), we obtain

$$p_l^n(1/2) = (-1)^{j-1/2}\sqrt{\frac{(n-l)(n+l+1)}{(n+1)(2l+1)}}C_{0,\,0,\,0}^{j-1/2,j-1/2,l} \ , \ n \text{ odd}, \ l \text{ even},$$

$$= (-1)^{j-1/2}\sqrt{\frac{l(l+1)}{(n+1)(2l+1)}}C_{0,\,0,\,0}^{j+1/2,j-1/2,l} \ , \ n \text{ odd}, \ l \text{ odd},$$

$$= (-1)^{j}\sqrt{\frac{n+1}{2l+1}}\left(1-\frac{l(l+1)}{n(n+2)}\right)C_{0,0,0}^{j,j,l} \ , \ n \text{ even}, \ l \text{ even}, \qquad (10.39)$$

$$= (-1)^{j-1}\sqrt{\frac{n+1}{2l+1}}\frac{\sqrt{(n-l)(n+l+1)l(l+1)}}{n(n+2)}C_{0,\,0,\,0}^{j,j-1,l} \ , \ n \text{ even}, \ l \text{ odd}.$$

Thus we observe by direct inspection that $p_l^n(1/2) \neq 0$ in all possible cases, being always a non-zero multiple of a Clebsch-Gordan coefficient of type $C_{0,0,0}^{l_1,l_2,l_3}$, with $(l_1 + l_2 + l_3)$ even.

Finally, Eqs. (6.52)–(6.55) follow straightforwardly from (10.39) and Eq. (4.49) for the Clebsch-Gordan coefficient of type $C_{0,0,0}^{l_1,l_2,l_3}$.

10.7 A Proof of Proposition 7.1.6

Proposition 7.1.6 follows from (7.10). But it can be proved more directly, without resorting to the formulas for the Wigner product symbol, as follows.

From the identities (4.30) we deduce the following formulae:

$$x = \frac{-1}{\sqrt{6}}(Y_{1,1} - Y_{1,-1}), y = \frac{i}{\sqrt{6}}(Y_{1,1} + Y_{1,-1}), z = \frac{1}{\sqrt{3}}Y_{1,0},$$

$$(x \pm iy)^2 = \sqrt{\frac{8}{15}}Y_{2,\pm2}, xy = \frac{-i}{\sqrt{30}}(Y_{2,2} - Y_{2,-2}), x^2 - y^2 = \sqrt{\frac{2}{15}}(Y_{2,2} + Y_{2,-2}).$$

Therefore, recalling the definition of the coupled standard basis $\{e(l,m)\}$, the symbol correspondence

$$W_1 : \mu_0 e(l,m) \longleftrightarrow Y_{l,m}$$

yields the specific correspondences

$$x \longleftrightarrow \frac{1}{\sqrt{6}}\frac{\mu_0}{\mu_1}(A+B), y = \frac{i}{\sqrt{6}}\frac{\mu_0}{\mu_1}(-A+B), z = \frac{1}{\sqrt{6}}\frac{\mu_0}{\mu_1}(AB-BA),$$

$$xy \longleftrightarrow \frac{-i}{\sqrt{30}}\frac{\mu_0}{\mu_2}(A^2 - B^2), x^2 - y^2 \longleftrightarrow \sqrt{\frac{2}{15}}\frac{\mu_0}{\mu_2}(A^2+B^2), \qquad (10.40)$$

where we have used the notation $A = J_+$, $B = J_-$. Consequently,

$$x \star_1^n y \longleftrightarrow \frac{1}{\sqrt{6}} \frac{\mu_0}{\mu_1}(A+B) \frac{i}{\sqrt{6}} \frac{\mu_0}{\mu_1}(-A+B)$$

$$= \frac{\sqrt{30}\mu_2}{n(n+2)\mu_0} \frac{-i}{\sqrt{30}} \frac{\mu_0}{\mu_2}(A^2-B^2) + \frac{i}{\sqrt{n(n+2)}} \frac{1}{\sqrt{6}} \frac{\mu_0}{\mu_1}(AB-BA)$$

$$\longleftrightarrow \frac{\sqrt{30}\mu_2}{n(n+2)\mu_0}(xy) + \frac{i}{\sqrt{n(n+2)}}z.$$

This gives the product formula (7.4) for $(x,y,z) = (a,b,c)$, and similarly one verifies the formula for a cyclic permutation of the coordinate functions.

On the other hand, by (3.14)–(3.15) we also have

$$x \longleftrightarrow \frac{2}{\sqrt{6}} \frac{\mu_0}{\mu_1} J_1, \ y \longleftrightarrow \frac{2}{\sqrt{6}} \frac{\mu_0}{\mu_1} J_2, \ z \longleftrightarrow \frac{2}{\sqrt{6}} \frac{\mu_0}{\mu_1} J_3, \tag{10.41}$$

which by (3.3) yields

$$x \star_1^n x + y \star_1^n y + z \star_1^n z \longleftrightarrow \frac{4}{n(n+2)}(J_1^2 + J_2^2 + J_3^2) = I,$$

and this proves the third identity (7.6).

Finally, let us calculate the three products $a * a$, for $a = x, y, z$, from three linear equations relating them. To this end, we start with the correspondences

$$A \longleftrightarrow \frac{\mu_1}{\mu_0} \sqrt{\frac{3}{2}}(x+iy), \ B \longleftrightarrow \frac{\mu_1}{\mu_0} \sqrt{\frac{3}{2}}(x-iy)$$

which yield

$$A^2 + B^2 \longleftrightarrow \frac{3\mu_1^2}{\mu_0^2}(x \star_1^n x - y \star_1^n y).$$

Combining this with (10.40) we obtain the identity

$$x \star_1^n x - y \star_1^n y = \frac{1}{3} \sqrt{\frac{15}{2} \frac{\mu_0 \mu_2}{\mu_1^2}}(x^2 - y^2),$$

and similarly, there is the identity

$$y \star_1^n y - z \star_1^n z = \frac{1}{3} \sqrt{\frac{15}{2} \frac{\mu_0 \mu_2}{\mu_1^2}}(y^2 - z^2).$$

These two equations together with Eq. (7.6) yield the solution (7.5).

10.8   A Proof of Proposition 7.2.28

In order to keep par with the convention used in [92], in this appendix we assume the Hermitian inner product $h_{n+1}(\cdot,\cdot) = <\cdot,\cdot>$ on $\mathcal{H}_j \equiv \mathbb{C}^{n+1}$, as well as all other inner products, to be conjugate linear in the second entry, not the first. Then, the standard Berezin symbol of an operator $T : \mathcal{H}_j \to \mathcal{H}_j$ is given by

$$B_T(\mathbf{n}) = h_{n+1}(T\tilde{Z}, \tilde{Z}) = <T\tilde{Z}, \tilde{Z}>$$

where

$$\mathbf{z} = (z_1, z_2) \in SU(2) = S^3 \subset \mathbb{C}^2, \ \Phi_j(\mathbf{z}) = \tilde{Z} \in \mathcal{M}_j \subset S^{2n+1} \subset \mathbb{C}^{n+1},$$

$$\Phi_j(\mathbf{z}) = \tilde{Z} = \left(z_1^n, \sqrt{\binom{n}{1}} z_1^{n-1} z_2, \ldots, \sqrt{\binom{n}{k}} z_1^{n-k} z_2^k, \ldots, z_2^n \right) \tag{10.42}$$

and $\mathbf{n} = \pi(\mathbf{z}) = [z_1, z_2] \in S^2$, π being the projection in the Hopf fibration

$$S^1 \to S^3 \to S^2 , \ \pi : S^3 \to S^2 .$$

The map $\Phi_j : S^3 \to S^{2n+1}$ is $SU(2)$-equivariant and is an embedding if j is half-integral, in which case its image is the orbit $\mathcal{M}_j \simeq SU(2) \simeq S^3$, whereas in the case of integral j the orbit is a manifold $\mathcal{M}_j \simeq SO(3) \simeq P^3$. To the Hopf fibration there is a related S^1 principal fibre bundle depending on j,

$$S^1 \to \mathcal{M}_j \to S^2 , \ \pi_j : \mathcal{M}_j \to S^2 , \ \pi = \pi_j \circ \Phi_j . \tag{10.43}$$

In what follows, it is important to highlight the explicit relation between the Hermitian metrics $h_{n+1} : \mathcal{M}_j \times \mathcal{M}_j \to \mathbb{C}$ and $h_2 : S^3 \times S^3 \to \mathbb{C}$ that is immediate from the explicit expression (10.42) of the map Φ_j, namely

$$h_{n+1}(\tilde{Z}, \tilde{Z}') = h_{n+1}(\Phi_j(\mathbf{z}), \Phi_j(\mathbf{z}')) = (h_2(\mathbf{z}, \mathbf{z}'))^n . \tag{10.44}$$

To simplify and keep close to the notation in [92], we shall denote points $\tilde{Z}, \tilde{Z}_1, \ldots$ on the orbit $\mathcal{M}_j$ by $m, m_1, m_2, \ldots$ and general vectors in $\mathbb{C}^{n+1}$ by $v, v_1, v_2, w, \ldots$.

Now, the manifold $\mathcal{M}_j$ inherits from the Hilbert space $\mathcal{H}_j$, viewed as a euclidean space $\mathbb{R}^{2n+2}$, an $SU(2)$-invariant Riemannian metric, and hence an $SU(2)$-invariant measure dm as well as an invariant L^2-inner product

$$<f, g>_{\mathcal{M}_j} := \int_{\mathcal{M}_j} f(m)\overline{g(m)}dm. \tag{10.45}$$

Thus, the idea set forth in [92] is to work out most of what is related to the standard Berezin correspondence at the level of the orbit $\mathcal{M}_j$, which is possible because of the explicit use of the Hermitian structure for this correspondence, as follows.

First, we may and shall assume the measure dm is "normalized" so that

$$v = \int_{\mathcal{M}_j} <v, m> m \, dm, \quad \text{for all } v \in \mathbb{C}^{n+1}. \tag{10.46}$$

In particular, this implies

$$< v_1, v_2 > = \int_{\mathcal{M}_j} <v_1, m><m, v_2> dm. \tag{10.47}$$

Lemma 10.8.1. *For any $T \in M_{n+1}(\mathbb{C})$,*

$$trace(T) = \int_{\mathcal{M}_j} < Tm, m > dm. \tag{10.48}$$

Proof. Let $\{e_i, i = 1, 2, \ldots n + 1\}$ be an orthonormal basis of $\mathbb{C}^{n+1}$, and hence

$$\forall m \in \mathcal{M}_j, \ m = \sum_{i=1}^{n+1} < m, e_i > e_i \ , \ \text{and } e_i = \int_{\mathcal{M}_j} < e_i, m > m \, dm \ ,$$

$$trace(T) = \sum_i < Te_i, e_i > \ = \sum_i < T(\int_{\mathcal{M}_j} < e_i, m > m \, dm, e_i >$$

$$= \sum_i < \int_{\mathcal{M}_j} <e_i, m > Tm \, dm, e_i > = \sum_i \int_{\mathcal{M}_j} < Tm, \overline{<e_i, m>} e_i > dm$$

$$= \int_{\mathcal{M}_j} < Tm, \sum_i < m, e_i > e_i > dm = \int_{\mathcal{M}_j} < Tm, m > dm.$$

$$\square$$

By choosing $T = Id$ in the above lemma we deduce the following:

$$Vol(\mathcal{M}_j) = \int_{\mathcal{M}_j} dm = n + 1 \ . \tag{10.49}$$

For any operator T, following [92] we define the function

$$K_T : \mathcal{H}_j \times \mathcal{H}_j \to \mathbb{C}, \ \ K_T(v, w) = \ < Tv, w > \tag{10.50}$$

and express T as an integral operator by integration over $\mathcal{M}_j$:

$$Tv = \int_{\mathcal{M}_j} K_T(v, m) m \, dm. \tag{10.51}$$

The validity of identity (10.51) follows from the normalization (10.46) of the measure dm. In view of (10.51), K_T is the integral kernel of T.

Clearly, the kernel of the composition $T_2 T_1$, as a function $\mathcal{H}_j \times \mathcal{H}_j \to \mathbb{C}$, can be expressed directly via an inner product using (10.50), but also as an integral

$$K_{T_2 T_1}(v, w) = \ < T_2 T_1 v, w > \ = \int_{\mathcal{M}_j} K_{T_1}(v, m) K_{T_2}(m, w) \, dm. \tag{10.52}$$

Now, we recall that for the standard Berezin correspondence determined by characteristic numbers $\vec{b}$, the covariant-to-contravariant transition operator on symbols

$$U^j_{\vec{b}, \frac{1}{\vec{b}}} : Poly_{\mathbb{C}}(S^2)_{\leq n} \to Poly_{\mathbb{C}}(S^2)_{\leq n} \ , \ f \mapsto \tilde{f}$$

corresponds to the transition operator (cf. Definition 7.1.18)

$$V^j_{\vec{b}, \frac{1}{\vec{b}}} : M_{\mathbb{C}}(n+1) \to M_{\mathbb{C}}(n+1) \ , \ F \mapsto F'$$

in such a way that

$$f = W^j_{\vec{b}}(F) = B_F \iff \tilde{f} = W^j_{\vec{b}}(F') = B_{F'} . \tag{10.53}$$

To keep par with the notation used in [92], we shall denote these transition operators and their respective inverses, as follows:

$$\eta^{-1} \equiv V^j_{\vec{b}, \frac{1}{\vec{b}}} \ , \ \eta \equiv V^j_{\frac{1}{\vec{b}}, \vec{b}} \ \ , \ \ \tilde{\eta}^{-1} \equiv U^j_{\vec{b}, \frac{1}{\vec{b}}} \ , \ \tilde{\eta} \equiv U^j_{\frac{1}{\vec{b}}, \vec{b}} \ ,$$

According to Eqs. (6.25) and (7.3), denoting the induced inner product on $Poly_{\mathbb{C}}(S^2)_{\leq n}$ by $< \cdot, \cdot >_{*\frac{n}{\vec{b}}}$ and the (usual) L^2-inner product on $C^\infty(S^2)$ by $< \cdot, \cdot >$,

$$\frac{1}{n+1} trace(FG^*) = <f, g>_{*\frac{n}{\vec{b}}} = <\tilde{f}, g> = <\tilde{\eta}^{-1}(B_F), g> = <B_{\eta^{-1}(F)}, B_G> ,$$

$$\frac{1}{n+1} trace(\eta(F)G^*) = < B_{\eta(F)}, g >_{*\frac{n}{\vec{b}}} = < \tilde{\eta}(f), g >_{*\frac{n}{\vec{b}}} \tag{10.54}$$

$$= \ < f, g > = < B_F, B_G > .$$

In what follows, we denote by σ_T the trivial lift of a standard Berezin symbol B_T on S^2 to the orbit $\mathcal{M}_j$, namely

$$\sigma_T(m) = B_T(\pi_j(m)).$$

Lemma 10.8.2. *For* $T \in M_{\mathbb{C}}(n+1)$, *the operator* $\eta(T)$ *can be expressed as follows:*

$$\eta(T)(v) = \int_{\mathcal{M}_j} B_T(\pi_j(m)) < v, m > m \, dm. \tag{10.55}$$

Proof. Define the operator $T_1 \in M_{\mathbb{C}}(n+1)$ by the right side of (10.55), namely

$$T_1(v) = \int_{\mathcal{M}_j} \sigma_T(m) < v, m > m \, dm. \tag{10.56}$$

We need to show that $T_1 = \eta(T)$. Now, $K_{T_1}(v,w) = < T_1 v, w >$, thus, by (10.56),

$$K_{T_1}(v,w) = \int_{\mathcal{M}_j} \sigma_T(m) < v, m > < m, w > dm.$$

But, given another operator S, the composition $T_1 S^*$ has by (10.52) kernel

$$K_{T_1 S^*}(v,w) = \int_{\mathcal{M}_j} K_{S^*}(v,m) K_{T_1}(m,w) dm \tag{10.57}$$

$$= \int_{\mathcal{M}_j} \int_{\mathcal{M}_j} < S^* v, m > \sigma_T(m') < m, m' > < m', w > dm' dm$$

$$= \int_{\mathcal{M}_j} [\int_{\mathcal{M}_j} < v, Sm > < m, m' > dm] \sigma_T(m') < m', w > dm'$$

$$= \int_{\mathcal{M}_j} < v, Sm' > \sigma_T(m') < m', w > dm'$$

where in the last step we have applied (10.47). Now, by (10.48), (10.57), and applying (10.47) again, we obtain

$$trace(T_1 S^*) = \int_{\mathcal{M}_j} K_{T_1 S^*}(m, m) dm \tag{10.58}$$

$$= \int_{\mathcal{M}_j} \int_{\mathcal{M}_j} \sigma_T(m') < m, Sm' > < m', m > dm' dm$$

$$= \int_{\mathcal{M}_j} \sigma_T(m') < m', Sm' > dm' = \int_{\mathcal{M}_j} \sigma_T(m') \overline{< Sm', m' >} dm'$$

$$= \int_{\mathcal{M}_j} \sigma_T(m')\overline{\sigma_S(m')} = < \sigma_T, \sigma_S >_{\mathcal{M}_j} = (n+1) < B_T, B_S >_{S^2}$$

$$\Rightarrow \frac{1}{n+1} trace(T_1 S^*) = < B_T, B_S >, \tag{10.59}$$

and since S is arbitrary it follows from (10.54) that $T_1 = \eta(T)$.   □

Now, let us denote by ω_1, ω_2 the standard Berezin symbols of T_1, T_2 lifted up to the orbit $\mathcal{M}_j$. Then, denoting by $\omega_1 \star \omega_2$ the standard Berezin symbol of $T_1 T_2$ lifted to $\mathcal{M}_j$, we have the following result.

Lemma 10.8.3. *For Berezin symbols $\omega_i = \sigma_{T_i}, i = 1, 2,$*

$$\omega_1 \star \omega_2(m) = \iint_{\mathcal{M}_j \times \mathcal{M}_j} B_1'(m, m_2, m_1)\tilde{\omega}_1(m_1)\tilde{\omega}_2(m_2)dm_1 dm_2 \tag{10.60}$$

where $\tilde{\omega}_i = \tilde{\eta}^{-1}(\omega_i)$ is the contravariant Berezin symbol of T_i lifted to $\mathcal{M}_j$, and

$$B_1'(m, m_1, m_2) = < m, m_1 >< m_1, m_2 >< m_2, m > . \tag{10.61}$$

Proof. Start with

$$< T_2 T_1 v, v > = < T_1 v, T_2^* v > = \int_{\mathcal{M}} < T_1 v, m >< m, T_2^* v > dm$$

$$= \int_{\mathcal{M}_j} < T_1 v, m >< T_2 m, v > dm. \tag{10.62}$$

Using Lemma 10.8.2, let us choose $S = \eta^{-1}(T_1)$, so that

$$T_1 v = \int_{\mathcal{M}_j} \sigma_{\eta^{-1}(T_1)}(m_1) < v, m_1 > m_1 dm_1 = \int_{\mathcal{M}_j} < \eta^{-1}(T_1)m_1, m_1 >< v, m_1 > m_1 dm_1$$

$$\Rightarrow < T_1 v, m > = \int_{\mathcal{M}_j} < \eta^{-1}(T_1)m_1, m_1 >< v, m_1 >< m_1, m > dm_1$$

$$\Rightarrow < T_2 T_1 v, v > = \int_{\mathcal{M}_j} \{[\int_{\mathcal{M}_j} < \eta^{-1}(T_1)m_1, m_1 >< v, m_1 >< m_1, m > dm_1]$$

$$\cdot [\int_{\mathcal{M}_j} < \eta^{-1}(T_2)m_2, m_2 >< m, m_2 >< m_2, v > dm_2]\} dm$$

$$\Rightarrow < T_2 T_1 v, v > = \iint_{\mathcal{M}_j \times \mathcal{M}_j} < \eta^{-1}(T_1)m_1, m_1 >< \eta^{-1}(T_2)m_2, m_2 >$$

$$\cdot < v, m_1 >< m_1, m_2 >< m_2, v > dm_1 dm_2.$$

Now, we write $v = m$ and obtain the desired result.   □

Because B_1' clearly descends to the level of S^2 (cf. (10.42) and (10.43)), the following result is immediate (cf. Eq. (7.68)):

Corollary 10.8.4. *Set* $\pi_j(m_i) = \mathbf{n}_i \in S^2$, $i = 1, 2, 3$. *Then*,

$$\overline{B_1'(m_1, m_2, m_3)} = \left(\frac{4\pi}{n+1}\right)^2 \mathbb{T}_{\bar{b}}^j(\mathbf{n}_1, \mathbf{n}_2, \mathbf{n}_3), \tag{10.63}$$

where conjugation on the l.h.s. is necessary to account for the different conventions of Hermitian product used for defining B_1' *and* $\mathbb{T}_{\bar{b}}^j$ *(cf. Remark 3.1.2).*

Now, from Eqs. (10.44) and (10.61), we have immediately

$$B_1'(m_1, m_2, m_3) = (h_2(\mathbf{z}_1, \mathbf{z}_2)h_2(\mathbf{z}_2, \mathbf{z}_3)h_2(\mathbf{z}_3, \mathbf{z}_1))^n, \tag{10.64}$$

and by a straightforward computation using formula (2.31) for the Hopf map and the convention that h_2 is conjugate linear in the second entry, we finally get

$$h_2(\mathbf{z}_1, \mathbf{z}_2)h_2(\mathbf{z}_2, \mathbf{z}_3)h_2(\mathbf{z}_3, \mathbf{z}_1) \tag{10.65}$$

$$= \frac{1}{4}(1 + \mathbf{n}_1 \cdot \mathbf{n}_2 + \mathbf{n}_2 \cdot \mathbf{n}_3 + \mathbf{n}_3 \cdot \mathbf{n}_1 - i[\mathbf{n}_1, \mathbf{n}_2, \mathbf{n}_3]).$$

Equations (10.63)–(10.65) are equivalent to Eq. (7.84).

10.9 A Proof of Proposition 7.2.30

We use notations and conventions from Appendix 10.8. Proposition 7.2.30 is equivalent to Lemma 10.9.2 below, which is a consequence of the following lemma.

Lemma 10.9.1. *At the level of* $\mathcal{M}_j$, *the kernel of the contravariant-to-covariant symbol transformation* $\tilde{\eta}$ *for the standard Berezin correspondence is the function*

$$N'(m, m') = |<m, m'>|^2.$$

Proof. Let $\omega = \sigma_T$. The operator $\tilde{\eta}$ is defined by

$$\tilde{\eta}(\omega)(m) = \int_{\mathcal{M}_j} N'(m, m')\omega(m')dm'.$$

Then, by Lemma 10.8.2,

$$\tilde{\eta}(\omega)(m) = \sigma_{\eta(T)}(m) = \; < \eta(T)m, m >$$

$$= \int_{\mathcal{M}_j} \sigma_T(m') < m, m' >< m', m > dm'$$

$$= \int_{\mathcal{M}_j} \omega(m')[< m, m' >< m', m >]dm'$$

$$= \int_{\mathcal{M}_j} N'(m, m')\omega(m')dm'$$

where $N'(m, m') \; = \; < m, m' >< m', m > = \; | < m, m' > |^2$.   □

Clearly, N' descends to the level of S^2 and we have the following result.

Lemma 10.9.2. *Let* $\mathbf{n} = \pi_j(m), \mathbf{n}' = \pi_j(m') \in S^2$. *Recalling that* $\eta = V_{\frac{1}{b}, \vec{b}}$,

$$N'(m, m') = | < m, m' > |^2 = \left(\frac{1 + \mathbf{n} \cdot \mathbf{n}'}{2} \right)^n = \frac{4\pi}{n+1} \mathbb{U}_{\frac{1}{b}, \vec{b}}(\mathbf{n}, \mathbf{n}') . \qquad (10.66)$$

Proof. It follows immediately from Eq. (10.44) that

$$| < m, m' > |^2 = |h_{n+1}(\tilde{Z}, \tilde{Z}')|^2 = |(h_2(\mathbf{z}, \mathbf{z}'))^n|^2 = (|h_2(\mathbf{z}, \mathbf{z}')|^2)^n,$$

and by a straightforward computation using Eq. (2.31) for the Hopf map,

$$|h_2(\mathbf{z}, \mathbf{z}')|^2 = (1 + \mathbf{n} \cdot \mathbf{n}')/2 .$$

The last equality in (10.66) is immediate from the definitions.   □

10.10   A Proof of Theorem 8.1.1

Clearly, Eqs. (8.5)–(8.7) are equivalent, so here we will focus on the expansion of type (8.6) in inverse powers of $n + 1$, namely,

$$(-1)^{2j+m_3} \sqrt{\frac{(2j + 1)(2l_3 + 1)}{(2l_1 + 1)(2l_2 + 1)}} \begin{bmatrix} l_1 & l_2 & l_3 \\ m_1 & m_2 & -m_3 \end{bmatrix} [j] \qquad (10.67)$$

$$= C_{0,0,0}^{l_1, l_2, l_3} C_{m_1, m_2, m_3}^{l_1, l_2, l_3} + \frac{1}{n+1} C_{m_1, m_2, m_3}^{l_1, l_2, l_3} P(l_1, l_2, l_3) + O((n+1)^{-2}) \qquad (10.68)$$

for $n = 2j >> 1$, $l_1, l_2, l_3 << n$ (we emphasize that, in what follows, this is equivalent to letting $n \to \infty$ keeping l_1, l_2, l_3 finite).

The Wigner product symbol, as expressed in (119), decomposes as follows:

$$\begin{bmatrix} l_1 & l_2 & l_3 \\ m_1 & m_2 & -m_3 \end{bmatrix}[j]$$

$$= \sqrt{(2l_1+1)(2l_2+1)(2l_3+1)} \begin{pmatrix} l_1 & l_2 & l_3 \\ -m_1 & -m_2 & +m_3 \end{pmatrix} \begin{Bmatrix} l_1 & l_2 & l_3 \\ j & j & j \end{Bmatrix}$$

$$= \sqrt{(2l_1+1)(2l_2+1)(2l_3+1)}(-1)^{l_1+l_2+l_3} \begin{pmatrix} l_1 & l_2 & l_3 \\ m_1 & m_2 & -m_3 \end{pmatrix} \begin{Bmatrix} l_1 & l_2 & l_3 \\ j & j & j \end{Bmatrix}$$

$$= (-1)^{l_1+l_2+l_3} \sqrt{(2l_1+1)(2l_2+1)(2l_3+1)} \frac{(-1)^{-l_1+l_2-m_3}}{\sqrt{2l_3+1}} C^{l_1,l_2,l_3}_{m_1,m_2,m_3} \begin{Bmatrix} l_1 & l_2 & l_3 \\ j & j & j \end{Bmatrix}$$

$$= (-1)^{l_3-m_3} \sqrt{(2l_1+1)(2l_2+1)} C^{l_1,l_2,l_3}_{m_1,m_2,m_3} \begin{Bmatrix} l_1 & l_2 & l_3 \\ j & j & j \end{Bmatrix},$$

and consequently we can write the expression (10.67) as

$$(-1)^{n+m_3} \sqrt{\frac{(2j+1)(2l_3+1)}{(2l_1+1)(2l_2+1)}} \begin{bmatrix} l_1 & l_2 & l_3 \\ m_1 & m_2 & -m_3 \end{bmatrix}[j]$$

$$= C^{l_1,l_2,l_3}_{m_1,m_2,m_3} \Phi(l_1,l_2,l_3;n+1) \tag{10.69}$$

where the function Φ is defined by

$$\Phi(l_1,l_2,l_3;n+1) = (-1)^n (-1)^{l_3} \sqrt{n+1} \sqrt{2l_3+1} \begin{Bmatrix} l_1 & l_2 & l_3 \\ j & j & j \end{Bmatrix} . \tag{10.70}$$

Next, using the expression (123) for the 6j-symbol $\{\}$ on the right-hand side of (10.70), we can express the above function as

$$\Phi(l_1,l_2,l_3;n+1) = (-1)^{l_3} \sqrt{2l_3+1} l_1! l_2! \, l_3! \Delta(l_1,l_2,l_3) \Upsilon(l_1,l_2,l_3;n+1)$$

where Υ is the only function depending on n, namely,

$$\Upsilon(l_1,l_2,l_3;n+1) \tag{10.71}$$

$$= \sqrt{\frac{(n+1)\cdot(n-l_1)!(n-l_2)!(n-l_3)!}{(n+l_1+1)!(n+l_2+1)!(n+l_3+1)!}} \sum_k \frac{(-1)^k(n+1+k)!}{(n+k-l_1-l_2-l_3)!R(l_1,l_2,l_3;k)} ,$$

with summation index k assuming all integral values for which all factorial arguments in $(n+k-l_1-l_2-l_3)!R(l_1,l_2,l_3;k)$ are nonnegative, where

$$R(l_1, l_2, l_3; k) = \prod_{i=1}^{3}(k - l_i)! \prod_{i<j}^{3}(l_i + l_j - k)! \,, \text{cf. (3.109)}. \tag{10.72}$$

We note that, for $n \gg l_i$, the restriction on summation index k amounts to demanding all factorial arguments in $R(l_1, l_2, l_3; k)$ being nonnegative, namely

$$\max\{l_i\} \le k \le \min\{l_i + l_j\} \;. \tag{10.73}$$

Now, writing $\mu = n + 1$, $L = l_1 + l_2 + l_3$, we can re-express $\Upsilon(l_1, l_2, l_3; n + 1)$ as follows:

$$\Upsilon(l_1, l_2, l_3; \mu) = \mu^{1/2}\left[\prod_{i=1}^{3}(\mu - l_i)\ldots(\mu + l_i)\right]^{-1/2}$$

$$\cdot \sum_{k}\frac{(-1)^k(\mu + k)!}{(\mu + k - L - 1)! R(l_1, l_2, l_3; k)}$$

$$= \mu^{-L-1}\left[\prod_{i=1}^{3}\left(1 - \frac{l_i}{\mu}\right)\ldots\left(1 + \frac{l_i}{\mu}\right)\right]^{-1/2}\sum_{k}\frac{(-1)^k(\mu + k)!}{(\mu + k - L - 1)! R(l_1, l_2, l_3; k)}$$

$$= \left[\prod_{i=1}^{3}\prod_{p_i=0}^{l_i}\left(1 - \left(\frac{p_i}{\mu}\right)^2\right)\right]^{-1/2}\frac{1}{\mu^{L+1}}\sum_{k}\frac{(-1)^k(\mu + k)!}{(\mu + k - L - 1)! \, R(l_1, l_2, l_3; k)} \,.$$

$$\tag{10.74}$$

Note that the inverse square root factor in (10.74) expands as $1 + O(\mu^{-2})$. Therefore, the first two terms in powers of $1/\mu$ in the expansion of $\Upsilon(l_1, l_2, l_3; \mu)$ are given by the first two terms in powers of $1/\mu$ in the expansion of

$$\Psi(l_1, l_2, l_3; \mu) = \frac{1}{\mu^{L+1}}\sum_{k}\frac{(-1)^k(\mu + k)!}{(\mu + k - L - 1)! R(l_1, l_2, l_3; k)} \tag{10.75}$$

$$= \sum_{k}\frac{(-1)^k S(k, L; \mu)}{R(l_1, l_2, l_3; k)} \tag{10.76}$$

where we have written

$$S(k, L; \mu) = \left(1 + \frac{k}{\mu}\right)\left(1 + \frac{k - 1}{\mu}\right)\ldots\left(1 + \frac{k - L}{\mu}\right) \,. \tag{10.77}$$

Clearly, the latter expands when $\mu \to \infty$ as

$$S(k, L; \mu) = 1 + \frac{1}{\mu}\left[\frac{L+1}{2}(2k - L)\right] + O(\mu^{-2}).\tag{10.78}$$

Consequently, the asymptotic expansion of Υ begins as follows:

$$\Upsilon(l_1, l_2, l_3; \mu) = \sum_k \frac{(-1)^k}{R(l_1, l_2, l_3; k)}\tag{10.79}$$

$$+ \frac{1}{\mu}\frac{(l_1 + l_2 + l_3 + 1)}{2}\sum_k \frac{(-1)^k(2k - (l_1 + l_2 + l_3))}{R(l_1, l_2, l_3; k)}$$

$$+ O(\mu^{-2})$$

where the summation index k is subject to the constraint (10.73). Thus, the expression (10.67), presented as (10.69), has the asymptotic expansion

$$C^{l_1, l_2, l_3}_{m_1, m_2, m_3}\Phi(l_1, l_2, l_3; \mu) = C^{l_1, l_2, l_3}_{m_1, m_2, m_3}(-1)^{l_3}\sqrt{2l_3 + 1}\,l_1!l_2!\,l_3!\Delta(l_1, l_2, l_3)$$

$$\cdot \left\{\sum_k \frac{(-1)^k}{R(l_1, l_2, l_3; k)} + \frac{1}{\mu}\frac{(l_1 + l_2 + l_3 + 1)}{2}\right.$$

$$\left.\sum_k \frac{(-1)^k(2k - (l_1 + l_2 + l_3))}{R(l_1, l_2, l_3; k)} + O(\mu^{-2})\right\}$$

$$= C^{l_1, l_2, l_3}_{m_1, m_2, m_3}\Phi_0(l_1, l_2, l_3) + \frac{1}{\mu}C^{l_1, l_2, l_3}_{m_1, m_2, m_3}\Phi_1(l_1, l_2, l_3) + O(\mu^{-2})\tag{10.80}$$

where

$$\Phi_0(l_1, l_2, l_3) = (-1)^{l_3}\sqrt{2l_3 + 1}\,l_1!l_2!\,l_3!\Delta(l_1, l_2, l_3)\sum_k \frac{(-1)^k}{R(l_1, l_2, l_3; k)},\tag{10.81}$$

$$\Phi_1(l_1, l_2, l_3) = (-1)^{l_3}\sqrt{2l_3 + 1}\,l_1!l_2!\,l_3!\Delta(l_1, l_2, l_3)\tag{10.82}$$

$$\cdot \frac{(l_1 + l_2 + l_3 + 1)}{2}\sum_k \frac{(-1)^k(2k - (l_1 + l_2 + l_3))}{R(l_1, l_2, l_3; k)}.$$

Thus, the analysis of the first two terms in (10.80) amounts to a closer look at the above functions Φ_0 and Φ_1, and below we shall divide into two cases accordingly.

10.10.1   The 0th Order Term

Let us set

$$\Sigma_0(l_1, l_2, l_3) = \sum_k \frac{(-1)^k}{R(l_1, l_2, l_3; k)}, \tag{10.83}$$

where as before, the summation index runs over the string of nonnegative integers k as in (10.73). Then, for the zeroth order term, if we perform the change of variables $k \to L - k$ in the summation Σ_0, we obtain

$$\Sigma_0(l_1, l_2, l_3) = (-1)^L \Sigma_0(l_1, l_2, l_3) \tag{10.84}$$

because

$$R(l_1, l_2, l_3; k) = R(l_1, l_2, l_3, L - k) \tag{10.85}$$

which implies that

$$\Sigma_0 \equiv \Phi_0 \equiv 0, \quad \text{if} \quad L = l_1 + l_2 + l_3 \text{ is an odd number,} \tag{10.86}$$

so the expression (10.81) can be non-zero only when L is even.

Recall the summation in Eq. (3.47), which defines explicitly the Clebsch-Gordan coefficients,

$$C^{l_1, l_2, l_3}_{m_1, m_2, m_3} = \delta(m_1 + m_2, m_3) \sqrt{2l_3 + 1} \, \Delta(l_1, l_2, l_3) \, S^{l_1, \ l_2, \ l_3}_{m_1, m_2, m_3}$$

$$\cdot \sum_z \frac{(-1)^z}{z!(l_1+l_2-l_3-z)!(l_1-m_1-z)!(l_2+m_2-z)!(l_3-l_2+m_1+z)!(l_3-l_1-m_2+z)!}$$

with $\Delta(l_1, l_2, l_3)$ and $S^{l_1, \ l_2, \ l_3}_{m_1, m_2, m_3}$ given respectively by (3.48) and (3.49). Then, setting $m_1 = m_2 = 0$ and $z = k - l_3$, the summation above becomes

$$\sum_k \frac{(-1)^{k-l_3}}{(k-l_1)!(k-l_2)!(k-l_3)!(l_1+l_2-k)!(l_2+l_3-k)!(l_3+l_1-k)!}$$

$$= (-1)^{l_3} \sum_k \frac{(-1)^k}{R(l_1, l_2, l_3; k)} = (-1)^{l_3} \Sigma_0(l_1, l_2, l_3).$$

Thus we have the formula

$$C^{l_1, l_2, l_3}_{0, 0, 0} = (-1)^{l_3} \sqrt{2l_3 + 1} \Delta(l_1, l_2, l_3) l_1! l_2! l_3! \Sigma_0(l_1, l_2, l_3) \tag{10.87}$$

where the r.h.s. expression is the same as (10.81), consequently

$$\Phi_0(l_1, l_2, l_3) = C_{0,0,0}^{l_1, l_2, l_3} .$$

(10.88)

We remark that the symmetry of Clebsch-Gordan coefficients (cf. (4.50)), besides the above discussion, implies that the quantity (10.88) vanishes when L is odd. We also note that the coefficient (10.88) is given in closed form by Eq. (4.49) and, in view of Eq. (10.87), this is equivalent to the following closed formula for Σ_0 when $L = l_1 + l_2 + l_3$ is even, namely

$$\Sigma_0(l_1, l_2, l_3) = (-1)^{L/2} \frac{(\frac{L}{2})!}{l_1!(\frac{L}{2} - l_1)!\, l_2!(\frac{L}{2} - l_2)!\, l_3!(\frac{L}{2} - l_3)!} .$$

(10.89)

10.10.2   The 1st Order Term

Let us set

$$\Sigma_1(l_1, l_2, l_3) = \sum_k \frac{(-1)^k\, k}{R(l_1, l_2, l_3; k)} ,$$

(10.90)

with summation over nonnegative integers k as in (10.73). Now, using the symmetry (10.85), we calculate

$$\sum_k \frac{(-1)^k (L - k)}{R(l_1, l_2, l_3; k)} = (-1)^L \sum_k \frac{(-1)^{L-k}(L-k)}{R(l_1, l_2, l_3, L - k)} = (-1)^L \Sigma_1$$

$$= L \sum_k \frac{(-1)^k}{R(l_1, l_2, l_3; k)} - \sum_k \frac{(-1)^k\, k}{R(l_1, l_2, l_3; k)} = L\Sigma_0 - \Sigma_1$$

and deduce the identity

$$[1 + (-1)^L]\Sigma_1 = L\Sigma_0$$

(10.91)

and hence

$$L \text{ odd} \implies \Sigma_0(l_1, l_2, l_3) = 0 \quad (\text{cf. } (10.86)),$$

$$L \text{ even} \implies \Sigma_1(l_1, l_2, l_3) = \frac{L}{2} \Sigma_0(l_1, l_2, l_3).$$

(10.92)

Therefore, using (10.91), Φ_1 given by (10.82) can be re-expressed as

$$\Phi_1(l_1, l_2, l_3) = (-1)^{l_3} \sqrt{2l_3 + 1} \Delta(l_1, l_2, l_3) l_1! l_2! l_3! \frac{(L+1)}{2} \tag{10.93}$$

$$\cdot \left(\sum_k \frac{(-1)^k k}{R(l_1, l_2, l_3; k)} - \sum_k \frac{(-1)^k (L-k)}{R(l_1, l_2, l_3; k)} \right)$$

$$= (-1)^{l_3} \sqrt{2l_3 + 1} \Delta(l_1, l_2, l_3) l_1! l_2! l_3! \frac{(L+1)}{2} \left(\Sigma_1 - (-1)^L \Sigma_1 \right)$$

$$= (-1)^{l_3} \frac{1 + (-1)^{L+1}}{2} \sqrt{2l_3 + 1} \Delta(l_1, l_2, l_3) l_1! l_2! l_3! (L+1) \cdot \Sigma_1.$$

Corollary 10.10.1. *With Σ_1 defined as in (10.90) and $L = l_1 + l_2 + l_3$, we have that $\Phi_1(l_1, l_2, l_3) = 0$ if L is even, and*

$$\Phi_1(l_1, l_2, l_3) = [(-1)^{l_3} \sqrt{2l_3 + 1} \Delta(l_1, l_2, l_3) l_1! l_2! l_3! (L+1)] \cdot \Sigma_1(l_1, l_2, l_3)$$

if L is odd.

Now, just as Σ_0 has the closed formula given by (10.89), Σ_1 has a similar closed formula which we shall seek. Denote by $[x]$ the integer part of a positive rational number, namely the largest integer $\leq x$. Set

$$Q(l_1, l_2, l_3) = \frac{[\frac{L}{2}]!}{l_1!([\frac{L}{2}] - l_1)! \, l_2!([\frac{L}{2}] - l_2)! \, l_3!([\frac{L}{2}] - l_3)!}, \tag{10.94}$$

so that $\Sigma_0 \equiv (-1)^{L/2} Q$, when $L = l_1 + l_2 + l_3$ is even.

Proposition 10.10.2. *The functions Σ_1 and Q defined by (10.90) and (10.94), respectively, are proportional to each other. More precisely:*

$$\Sigma_1(l_1, l_2, l_3) = \frac{(-1)^{[\frac{L+1}{2}]}}{2} \left(1 + \left[\frac{L+1}{2} \right] + (-1)^L \left[\frac{L-1}{2} \right] \right) Q(l_1, l_2, l_3)$$

$$\tag{10.95}$$

$$= \begin{cases} (-1)^{\frac{L}{2}} \dfrac{L}{2} Q(l_1, l_2, l_3) \,, & \text{if } L \text{ is even} \\[2mm] (-1)^{\frac{L+1}{2}} Q(l_1, l_2, l_3) \,, & \text{if } L \text{ is odd.} \end{cases}$$

For L even, the above expression for Σ_1 is immediate from (10.89) and (10.92), namely it follows from simple symmetry considerations. For L odd, a formal (combinatorial) proof of the above formula for Σ_1 has so far eluded us. The reader may easily verify the formula with a computer program, say for odd $L = l_1 + l_2 + l_3$ up to $\simeq 40,000$, recalling that the numbers l_i are subject to the triangle inequalities $\delta(l_1, l_2, l_3) = 1$, cf. (3.36).

Therefore, from the equations in Corollary 10.10.1 and Proposition 4.2.8 we finally obtain that, regardless of whether L is even or odd, we always have

$$\Phi_1 \equiv P$$

and (cf. Eqs. (10.68) and (10.80)) this concludes the proof of Theorem 8.1.1.

Bibliography

1. R. Abraham, J.E. Marsden, *Foundations of Mechanics* (Benjamin, Reading, 1978)
2. J.F. Adams, *Lectures on Lie Groups* (Benjamin, New York/Amsterdam, 1969)
3. V. Aquilanti, H.M. Haggard, R.G. Littlejohn, L. Yu, Semiclassical analysis of Wigner $3j$-symbols. J. Phys. A: Math. Theor. **40**, 5637–5674 (2007)
4. V. Aquilanti, H.M. Haggard, A. Hedeman, N. Jeevangee, R.G. Littlejohn, L. Yu, Semiclassical mechanics of the Wigner $6j$-symbol. J. Phys. A: Math. Theor. **45**, 065209, 61 (2012)
5. V.I. Arnold, *Mathematical Methods of Classical Mechanics*. Graduate Texts in Mathematics, vol. 60 (Springer, New York/Berlin, 1989)
6. A. Baker, *Matrix Groups: An Introduction to Lie Group Theory* (Springer, New York/Berlin, 2003)
7. V. Bargmann, On the representations of the rotation group. Rev. Mod. Phys. **34**, 829–845 (1962)
8. F. Bayen, C. Frondsal, Quantization on the sphere. J. Math. Phys. **22**, 1345–1349 (1981)
9. F. Bayen, M. Flato, C. Frondsal, A. Lichnerowicz, D. Sternheimer, Deformation theory and quantization. Ann. Phys. **111**, 61–151 (1977)
10. F.A. Berezin, Quantization. Math. USSR Izvest. **8**, 1109–1163 (1974)
11. F.A. Berezin, Quantization in complex symmetric spaces. Math. USSR Izvest. **9**, 341–379 (1975)
12. F.A. Berezin, General concept of quantization. Commun. Math. Phys. **40**, 153–174 (1975)
13. L.C. Biedenharn, J.D. Louck, *Angular Momentum in Quantum Physics* (Addison-Wesley, Reading, 1981)
14. L.C. Biedenharn, J.D. Louck, *The Racah-Wigner Algebra in Quantum Theory* (Addison-Wesley, Reading, 1981)
15. P. Bieliavsky, S. Detournay, Ph. Spindel, The deformation quantizations of the hyperbolic plane. Commun. Math. Phys. **289**, 529–559 (2009)
16. A. Bohm, *Quantum Mechanics: Foundations and Applications* (Springer, New York/Berlin, 1993)
17. N. Bohr, On the constitution of atoms and molecules, Parts I and II. Philos. Mag. **26**, 1–24, 476–512 (1913)
18. M. Bordermann, E. Meinrenken, M. Schlichenmaier, Toeplitz quantization of Kähler manifolds and $gl(N)$, $N \to \infty$ limits. Commun. Math. Phys. **165**, 281–296 (1994)
19. M. Born, P. Jordan, Zur Quantenmechanik. Z. Phys. **34**, 858–888 (1925)
20. M. Born, W. Heisenberg, P. Jordan, Zur Quantenmechanik II. Z. Phys. **35**, 557–615 (1925)
21. L. Boutet de Monvel, V. Guillemin, *The Spectral Theory of Toeplitz Operators*. Annals of Mathematics Studies, vol. 99 (Princeton University Press, Princeton, 1981)

© Springer International Publishing Switzerland 2014   193
P. de M. Rios, E. Straume, *Symbol Correspondences for Spin Systems*,
DOI 10.1007/978-3-319-08198-4

22. P.J. Brussard, H.A. Tolhoek, Classical limits of Clebsch-Gordan coefficients, Racah coefficients and $D^l_{mn}(\varphi, \vartheta, \psi)$-functions. Physica **23**, 955–971 (1957)
23. C. Cohen-Tannoudji, B. Diu, F. Laloe, *Quantum Mechanics*, vols. 1 and 2 (Hermann, Paris/Wiley, New York, 1977)
24. E.U. Condon, Q.W. Shortley, *The Theory of Atomic Spectra* (Cambridge University Press, Cambridge, 1935)
25. P.A.M. Dirac, The fundamental equations of quantum mechanics. Proc. R. Soc. Lond. A **109**, 642–653 (1925)
26. P.A.M. Dirac, *The Principles of Quantum Mechanics* (Oxford University Press, Oxford, 1958)
27. J.J. Duistermaat, L. Hörmander, Fourier integral operators II. Acta Math. **128**, 183–269 (1972)
28. J.J. Duistermaat, J.A.C. Kolk, *Lie Groups* (Springer, New York/Berlin, 2000)
29. B.V. Fedosov, A simple geometrical construction of deformation quantization. J. Differ. Geom. **40**(2), 213–238 (1994)
30. L. Freidel, K. Krasnov, The fuzzy sphere *-product and spin networks. J. Math. Phys. **43**(4), 1737–1754 (2002)
31. I.M. Gelfand, N.J. Vilenkin, *Generalized Functions, Volume 4: Some Applications of Harmonic Analysis. Rigged Hilbert Spaces* (Academic, New York, 1964)
32. R. Gilmore, *Lie Groups, Physics and Geometry* (Cambridge University Press, Cambridge, 2008)
33. R.J. Glauber, Coherent and incoherent states of radiation field. Phys. Rev. **131**, 2766–2788 (1963)
34. H. Goldstein, C. Poole, J. Safko, *Classical Mechanics*, 3rd edn. (Addison-Wesley, New York, 2001)
35. H.J. Groenewold, On the principles of elementary quantum mechanics. Physica **12**, 405–460 (1946)
36. A. Grossmann, G. Loupias, E.M. Stein, An algebra of pseudo-differential operators and quantum mechanics in phase space. Ann. Inst. Fourier **18**, 343–368 (1968)
37. V. Guillemin, S. Sternberg, *Symplectic Techniques in Physics* (Cambridge University Press, Cambridge/New York, 1984)
38. W. Heisenberg, Uber quantentheoretische Umdeutung kinematischer und mechanischer Beziehungen. Z. Phys. **33**, 879–893 (1925)
39. L. Hörmander, Pseudo-differential operators. Commun. Pure Appl. Math. **18**, 501–517 (1965)
40. L. Hörmander, Fourier integral operators I. Acta Math. **127**, 79–183 (1971)
41. A.W. Knapp, *Lie Groups Beyond an Introduction*. Progress in Mathematics, vol. 140, 2nd edn. (Birkhauser, Boston/Berlin, 2002)
42. M. Kontsevich, Deformation quantization of Poisson manifolds. Lett. Math. Phys. **66**(3), 157–216 (2003)
43. A. Koyré, Galileo and Plato. J. Hist. Ideas **4**(4), 400–428 (1943)
44. N.P. Landsman, Strict quantization of coadjoint orbits. J. Math. Phys. **39**(12), 6372–6383 (1998)
45. R.G. Littlejohn, The semiclassical evolution of wave packets. Phys. Rep. **138**, 193–291 (1986)
46. R.G. Littlejohn, *Rotations in Quantum Mechanics, and Rotations of Spin-$\frac{1}{2}$ Systems*. Lecture Notes in Physics #10 (University of California, Berkeley, 1996)
47. J. Madore, The fuzzy sphere. Class. Quantum Gravity **9**, 69–87 (1992)
48. J.E. Marsden, T.S. Ratiu, *Introduction to Mechanics and Symmetry* (Springer, New York/Berlin, 1999)
49. D. McDuff, D. Salamon, *Introduction to Symplectic Topology* (Oxford University Press, New York, 1998)
50. J.E. Moyal, Quantum mechanics as a statistical theory. Proc. Camb. Philos. Soc. **45**, 99–124 (1949)
51. A.M. Ozorio de Almeida, The Weyl representation in classical and quantum mechanics. Phys. Rep. **295**, 265–342 (1998)
52. W. Pauli, Zur Frage der Theoretischen Deutung der Satelliten einiger Spektrallinen und ihrer Beeinflussung durch magnetische Felder. Naturwissenschaften **12**(37), 741–743 (1924)

53. A. Perelomov, *Generalized Coherent States and Their Applications* (Springer, Berlin, 1986)
54. L. Polterovich, *The Geometry of the Group of Symplectic Diffeomorphisms* (Springer, Basel, 2001)
55. G. Ponzano, T. Regge, Semiclassical limit of Racah coefficients, in *Spectroscopic and Group Theoretical Methods in Physics*, ed. by F. Bloch, S.G. Cohen (North Holland, Amsterdam, 1968)
56. M. Rieffel, Deformation quantization of Heisenberg manifolds. Commun. Math. Phys. **122**, 531–562 (1989)
57. M. Rieffel, Matrix algebras converge to the sphere for quantum Gromov-Hausdorff distance. Mem. Am. Math. Soc. **168**(796), 67–91 (2004)
58. M. Rieffel, Leibniz seminorms for "matrix algebras converge to the sphere". Quanta of maths (Clay Math. Proc.) **11**, 543–578 (2010)
59. P. de M. Rios, A semiclassically entangled puzzle. J. Phys. A: Math. Theor. **40**, F1047–1052 (2007)
60. P. de M. Rios, A. Ozorio de Almeida, On the propagation of semiclassical Wigner functions. J. Phys. A: Math. Gen. **35**, 2609–2617 (2002)
61. P. de M. Rios, A. Ozorio de Almeida, A variational principle for actions on symmetric symplectic spaces. J. Geom. Phys. **51**, 404–441 (2004)
62. P. de M. Rios, G.M. Tuynman, Weyl quantization from geometric quantization. A.I.P. Conf. Proc. **1079**, 26–38 (2008)
63. M.E. Rose, *Elementary Theory of Angular Momentum* (Wiley, New York/Chapman & Hall, Ltd., London, 1957)
64. A. Royer, Wigner function as the expectation value of a parity operator. Phys. Rev. A **15**, 449–450 (1977)
65. J.J. Sakurai, *Modern Quantum Mechanics* (Addison-Wesley, New York, 1994)
66. E. Schrödinger, Quantisierung als Eigenwertproblem. Ann. der Physik **386**, 109–139 (1926)
67. R.L. Stratonovich, On distributions in representation space. Sov. Phys. JETP **31**, 1012–1020 (1956)
68. E. Straume, Weyl groups and the regularity properties of certain compact Lie group actions. Trans. AMS **306**, 165–190 (1988)
69. E. Straume, *Lecture Notes on Lie Groups and Lie Algebras* (NTNU, Trondheim, 1998)
70. E.C.G. Sudarshan, Equivalence of semiclassical and quantum mechanical descriptions of statistical light beams. Phys. Rev. Lett. **10**, 277–279 (1963)
71. G. Uhlenbeck, S. Goudsmit, Ersetzung der Hypothese vom unmechanischen Zwang durch eine Forderung bezuglich des inneren Verhaltens jedes einzelnen Elektrons. Naturwissenschaften **13**(47), 953–954 (1925)
72. G. Uhlenbeck, S. Goudsmit, Spinning electrons and the structure of spectra. Nature **117**, 264–265 (1926)
73. J.C. Várilly, J.M. Gracia-Bondía, The Moyal representation of spin. Ann. Phys. **190**, 107–148 (1989)
74. D.A. Varshalovich, A.N. Moskalev, V.K. Khersonskii, *Quantum Theory of Angular Momentum* (World Scientific, Singapore, 1988)
75. N.J. Vilenkin, A.U. Klimyk, *Representations of Lie Groups and Special Functions*, vol. 1 (Kluwer Academic, Dordrecht/Boston, 1991)
76. J. von Neumann *Mathematische Begrundung der Quantenmechanik*. Nachrichten Gottinger Gesellschaft der Wissenschaften, 1–57 (1927)
77. J. von Neumann, Die Eindeutigkeit der Schrödingerschen Operatoren. Math. Ann. **104**, 570–578 (1931)
78. J. von Neumann, *Mathematical Foundations of Quantum Mechanics* (Princeton University Press, Princeton, 1966). Translated from German original of 1932
79. A. Voros, An algebra of pseudo-differential operators and the asymptotics of quantum mechanics. J. Funct. Anal. **29**, 104–132 (1978)
80. A. Wassermann, Ergodic actions of compact groups on operator algebras. III Classification for $SU(2)$. Invent. Math. **93**, 309–354 (1988)

81. A. Weil, Sur certains groupes d'operateurs unitaires. Acta Math. **111**, 143–211 (1964)
82. A. Weinstein, The local structure of Poisson manifolds. J. Differ. Geom. **18**, 523–557 (1983)
83. A. Weinstein, Traces and triangles in symplectic symmetric spaces. Contemp. Math. **179**, 261–270 (1994)
84. A. Weinstein, Groupoids: unifying internal and external symmetry. A tour through some examples. Not. Am. Math. Soc. **43**, 744–752 (1996)
85. H. Weyl, Quantenmechanik und Gruppentheorie. Z. Phys. **46**, 1–46 (1927)
86. H. Weyl, *The Theory of Groups and Quantum Mechanics* (Dover, New York, 1931). Translated from German original of 1928
87. H. Weyl, *The Classical Groups. Their Invariants and Representations* (Princeton University Press, Princeton, 1939)
88. E. Wigner, Einige Folgerungen aus der Schrödingerschen Theorie fur die Termstrukturen. Z. Phys. **43**, 624–652 (1927)
89. E. Wigner, Berichtigung zu der Arbeit: Einige Folgerungen aus der Schrödingerschen Theorie fur die Termstrukturen. Z. Phys. **45**, 601–602 (1927)
90. E. Wigner, On the quantum correction for thermodynamic equilibrium. Phys. Rev. **40**, 749–759 (1932)
91. E. Wigner, *Group Theory and Its Application to the Quantum Mechanics of Atomic Spectra* (Academic, New York, 1959). Translated from German original of 1931
92. N.J. Wildberger, On the Fourier transform of a compact semisimple Lie group. J. Aust. Math. Soc. Ser. A **56**(1), 64–116 (1994)

Index

Adjoint
  matrix, 9
  representation, 15–16
Affine
  mechanical systems, 71–76
    Hilbert space, 73
  metaplectic group, 75
  symplectic form, 72
  symplectic group, 72
  symplectic space, 72
  translations, 72
Angular momentum (spin)
  commutation relations, 17
  operators, 16, 26–30
Anti-commutator, 111, 144
Associated Legendre polynomials, 65–66

Berezin symbol, 84. *see also* Symbol
      correspondences
Berezin transform, 129. *see also* Special
      functional transforms
Bohr's correspondence principle, 59
Bohr type, 151. *see also* Symbol
      correspondence sequences

Clebsch–Gordan coefficients, 34–36
  explicit formulae, 35
  non-vanishing conditions, 34
  orthogonality equations, 34
  relation with Wigner 3*jm* symbols, 48
  symmetry properties, 36
Coadjoint
  orbit, 16, 20, 59

representation, 15–16
symplectic form, 16
Coherent states, 99. *see also* States

Dequantization, 75
Dequantization groupoid, 113
Deviation, 104

Expectation, 100
Exponential map, 13

Groenewold-von Neumann
  product, 113
  trikernel, 4
Groups
  action, 8
  affine metaplectic, 75
  affine symplectic (real), 72
  definition, 7
  Heisenberg group, 73
  homomorphism, 7
  isomorphism, 8
  linear symplectic (real), 72
  metaplectic, 75
  orthogonal, 9
  subgroup, 7
  symplectic (quaternionic), 9
  unitary, 9
  unitary representations, 10, 21, 25, 73

Haar measure, 11
Hamilton-Poisson dynamics, 58

© Springer International Publishing Switzerland 2014
P. de M. Rios, E. Straume, *Symbol Correspondences for Spin Systems*,
DOI 10.1007/978-3-319-08198-4

Printed in the United States
By Bookmasters